GAS, DUST AND HYBRID EXPLOSIONS

FUNDAMENTAL STUDIES IN ENGINEERING

Vol. 1 Coelho/Physics of Dielectrics for the Engineer
Vol. 2 Szücs/Similitude and Modelling
Vol. 3 Scharf, Lisieski/Amplitude Distribution Spectrometers
Vol. 4 Mohilla, Ferenz/Chemical Process Dynamics
Vol. 5 Baker et al./Explosion Hazards and Evaluation
Vol. 6 Sobczyk/Stochastic Wave Propagation
Vol. 7 König/Shakedown of Elastic-Plastic Structures
Vol. 8 Cacko, Bily, Bukoveczky/Random Processes
Vol. 9 Kubácek/Foundations of Estimation Theory
Vol. 10 Catuneanu, Mihalache/Reliability Fundamentals
Vol. 11 Boros/Measurement Evaluation
Vol. 12 Baker et al./Similarity Methods in Engineering Dynamics, Revised Edition
Vol. 13 Baker, Tang/Gas, Dust and Hybrid Explosions

FUNDAMENTAL STUDIES IN ENGINEERING 13

GAS, DUST AND HYBRID EXPLOSIONS

Wilfred E. BAKER†
Wilfred Baker Engineering, Inc., San Antonio, Texas, U.S.A.

Ming Jun Tang
East China Institute of Technology, Nanjing, China

ELSEVIER
Amsterdam - Oxford - New York - Tokyo 1991

ELSEVIER SCIENCE PUBLISHERS B.V.
Sara Burgerhartstraat 25
P.O. Box 211, 1000 AE Amsterdam, The Netherlands

Distributors for the United States and Canada:

ELSEVIER SCIENCE PUBLISHING COMPANY INC.
655 Avenue of the Americas
New York, N.Y. 10010, U.S.A.

Library of Congress Cataloging-in-Publication Data

Baker, W. E.
Gas, dust, and hybrid explosions / Wilfred E. Baker, Ming Jun Tang.
p. cm. -- (Fundamental studies in engineering ; 13)
Includes bibliographical references and index.
ISBN 0-444-88150-6
1. Explosions. 2. Dust explosions. I. Tang, Ming Jun. II. Title. III. Series.
QD516.B18 1991
620.8'6--dc20
91-11576
CIP

ISBN: 0-444-88150-6

Printed in The Netherlands

PREFACE

The physics of explosions in air from condensed, high explosive sources such as TNT has been adequately discussed in the literature, including a book by one author of this text. But, many other classes of accidental explosion can and have occurred in a wide variety of industries and operations involving energetic materials. As was true about fifteen years ago for explosions of high explosive sources, much of the theory and experiment regarding explosions of lower energy density sources such as bursting pressure vessels, explosive gas mixtures, dust-air suspensions, liquid propellant mixtures, and hybrid mixtures is fragmented and can only be retrieved by a reader through careful study of a rather diverse literature. This book is an attempt to gather and present the available material for explosions in air of the "non-ideal" explosion sources most often encountered in industrial accidents.

Both theoretical and experimental approaches to these classes of explosions are covered, with rather complete descriptions of the physical processes known or believed to occur. Types of explosion sources include combustible dusts, chemically reactive gases, compressed but non-reactive gases and flash-evaporating fluids, liquid rocket propellants, and hybrid mixtures of combustible dusts and gases.

The approach is to consider each class of explosion source separately, first discussing fundamentals, then presenting methods of analysis and testing, and finally giving curves or equations to predict effects of the particular class of explosion.

An extensive bibliography is included. Tables of pertinent properties of potentially explosive materials are given in appendices.

The book deliberately omits discussions of reflected or diffracted blast wave parameters, and damaging effects of blasts. These topics are treated in detail in other texts and manuals.

It is hoped that this book will become a useful counterpart to Cole's Underwater Explosions and Baker's Explosions in Air.

W.E.B.
M.J.T.

ACKNOWLEDGEMENTS

Although the first author organized and wrote most of this book, he could not have completed it without considerable help. Most importantly, the second author, Professor Ming Jun Tang, East China Institute of Technology, Nanjing, China, wrote most of the important Chapter III on reactive gas explosions during and following a visit to San Antonio in 1989. She also provided a number of pertinent references for this and other topics. Her knowledge and expertise have greatly enhanced the book.

The author also gratefully acknowledges preparation of the final camera-ready manuscript by Lynette Ramon, Rosemary Rivas and Loretta Mesa, and preparation of many of the figures in the text by Victoriano Hernandez.

PUBLISHERS' NOTE

This book had just been taken into production when the grievous news was received of the death of its first author, the distinguished scholar in the field of explosion physics, Dr. Wilfred E. Baker.

Dr. Baker's outstanding contributions to the science of explosions and to the protection of human lives and properties from accidental explosions have long been highly regarded internationally. His publications, five books and numerous papers, have been widely used by both scientists and engineers throughout the world as basic guidelines to solve any problems on explosion phenomena. His brilliant dedication of all his energies and wisdom to the founding of explosion hazard evaluation as a branch of science will always be remembered by his readers worldwide.

This book is the last testimony to the unsparing talent of Dr. Baker. The Publishers are proud of having been instrumental in bringing his works to the attention of a wider audience.

TABLE OF CONTENTS

Section

Section

LIST OF TABLES

LIST OF FIGURES

Figure No.

Figure No.

Figure No.

Figure No.

Figure No.

Figure No.

Figure No.

I. INTRODUCTION

Accidental explosions have plagued industry almost from the start of the industrial revolution, and unfortunately they still occur more often than we would like. The majority of such accidents are probably relatively benign, causing no or little damage or injury, and at most startling people by the "bang" accompanying the explosion. But, a significant number are severe enough to cause property damage, injury, or death.

A prerequisite for any explosion is a rather rapid release of energy in or from a limited volume. This covers a wide spectrum of possible accidental explosion sources, which Strehlow and Baker (1976) have classified into the following groups:

1) Condensed phase sources
2) Combustion explosions in enclosures
3) Pressure vessel bursts
4) Boiling Liquid Expanding Vapor Explosions (BLEVEs)
5) Unconfined Vapor Cloud Explosions (UVCEs)
6) Physical vapor explosions
7) Nuclear reactor runaway

The first group contains primarily exothermic chemical materials which are designed to explode, or at least to release their energy quite rapidly. It incudes high explosives, solid propellants and mixed fuels and oxidizers of liquid propellants. In this book, we will not cover explosions of high explosives or solid propellants, because good references are already available [Baker (1973), Baker, et al (1983), Swisdak (1975)], but we will discuss explosions of liquid propellant mixtures because there has been extensive testing of such mixtures and little data are available in the open literature.

The second group includes a number of sources which are very important in industrial safety, and we include discussions of all of these sources. They are all enclosed mixtures of fuels and oxidants, with oxygen in the air usually being the oxidant. Dust and air, gas and air and fuel mist and air as explosive sources are covered, as well as "hybrid" mixtures of combustible dusts and gases with air.

The third group includes failures of pressure vessels containing compressed gases or flash-evaporating fluids, where the stored thermodynamic energy in the fluid is the explosion source. Again, this class of explosion source is covered quite thoroughly in this book.

The fourth group, with the acronym BLEVE, represents a safety hazard more from the aftereffects of fireballs and fires following the explosion, and from "rocketing" or hurling of large pieces of the failed pressure vessels. As explosion sources, they differ little from failures of a vessel containing a flash-evaporating fluid, so we do not include extensive discussions of blast waves accompanying BLEVEs.

Unlike BLEVEs, the fifth group (UVCEs), has as a major damaging effect from the explosion and accompanying blast wave. The occurrence of several truly catastrophic industrial explosions of this type have generated much related research and testing, which we summarize in this book.

In the foundry industry, the papermaking industry, and postulated for certain accident sequences for nuclear power plants, physical vapor explosions (or steam explosions) are a cause for concern. There being essentially no measurements or validated prediction methods for explosion effects for this group, we do <u>not</u> discuss them further in this book.

Finally, nuclear reactor runaway conditions have been postulated to generate a variety of dynamic pressure effects or chemical reactions leading to explosive mixtures. Again, we do not discuss those postulated explosions.

Dust explosions are discussed in Chapter II. We first present a discourse on fundamentals of the topic, then note the strong effect of confinement on dust explosions. Results of a similitude analysis giving scaling laws for vented dust explosions are then followed by a discussion of testing methods and explosion control methods. Because dust explosions in grain elevators represent probably the most severe and costly of this class of accidental explosion, we include a special discussion of that topic.

Chapter III covers all types of reactive gas explosions. It also starts with a discussion of fundamentals, including particularly the differences between burns and detonations. A general scaling law for this class of explosion is presented next. Then, we cover confined explosions, both deflagrations and detonations. The next topic is UVCEs, including properties of free-air blast waves generated by these explosions. Control methods are then discussed, and testing methods for assessing gas explosion hazards.

It is probable that most serious coal mine explosions are "hybrid" explosions, involving mixtures of both methane and coal dust with air. These very violent explosions, and others in other industries, highlight the enhanced hazards for this class of explosion. The same general approach as in earlier chapters is followed in Chapter IV in discussing this topic, with fundamentals presented first, followed by scaling, test methods, and control methods.

Chapter V covers explosions of systems which release stored pressure energy. Apart from explosions of high explosives, this is a group for which we have the best fund of information, both theoretical and experimental, to assess explosions and blast wave properties. The order of presentation is fundamentals, scaling, results of analysis and results of tests.

The last chapter, Chapter VI, covers the topic of liquid propellant explosions. This topic is important for safety considerations at rocket launch and handling facilities, and was the subject of intensive study by NASA and the U.S. Air Force in the 1960-1970 decade. The mixing of propellants and oxidizers, which is usually quite complex and heterogeneous for the common rocket propellant combinations, is discussed first, followed by a discussion of classes of explosions which can occur in launch pad accidents. Scaled blast data from past test work are also presented.

Throughout the book, we include curves or equations for scaled free-air blast parameters for those classes of accidental explosions where predictions are possible. The text material is supplemented by appendices and by an extensive bibliography.

II. DUST EXPLOSIONS

A. Fundamentals

Dusts are quantities of solids as small particles--small enough that they remain suspended in air for significant times when airborne. Field (1982) notes that particles with diameters up to 500μm can be airborne and can be considered as explosible dusts, if they are combustible.

When do combustible dusts explode? Most combustible dusts are fuels which can burn, but they can only burn if an oxidizer is present, and if they are ignited.[1]

In essentially all accelerated explosions involving dusts, significant confinement of the dust-oxidizer mixture is a prerequisite for an explosion. Unconfined dust clouds can be ignited, flame can propagate through the clouds, and large fireballs result. But purely combustible dust clouds apparently cannot be exploded, producing blast waves, from "soft" ignition by low-energy ignition sources.

Explosions are indeed possible for unconfined clouds of either some purely combustible dusts of high reactivity, e.g., flake aluminum powder, and of dusts of inherently explosive materials such as solid high explosives, with "hard" ignition from intense explosion sources such as high explosives.

There is a truly voluminous literature on control of dust explosions and their effects. Because a number of excellent comprehensive references are available, we will not try to duplicate that material and offer a competing treatise in this subject. Instead, we will summarize, highlighting our own experience and giving a guide to available literature. Other excellent comprehensive references on this topic, in addition to Field (1982) are Bartknecht (1978a), Bartknecht (1981), Palmer (1973), and Cashdollar and Hertzberg (1987). Related symposium proceedings include "Flammable Dust Explosions" (1988), "First Int. Colloquium on Explosibility of Industrial Dusts" (1984), and "Shenyang Int. Symp. on Dust Explosions" (1987).

B. The strong effect of confinement

As noted before, except in certain special cases with very strong ignition sources, dust explosions producing damaging pressures cannot occur without strong confinement. Explosibility of combustible dusts is determined experimentally

---

[1] Exceptions are dusts of solid explosives, solid propellants or similar materials, which contain both fuel and oxidizer within each particle.

by testing in unvented containment vessels of various sizes, as will be discussed in more detail later in this chapter. Such testing serves to establish upper limits on deflagrative explosion pressures P_m, and the maximum rates of rise of such pressures $(dP/dt)_{max}$ within "compact" enclosures, i.e., enclosures whose maximum internal principal dimensions are about equal (spheres, cubes, or short cylinders).

Typically, pressure variations with time within such unvented enclosures appear as shown in Fig. 2-1. Pressure initially increases quite slowly after dust ignition, following nearly a cubic increase with time. As combustion ends, the gage pressure peaks at a maximum value P_m, and then decreases because of heat losses to and through the test vessel. Although pressure variations of this character are measured for almost all combustible dusts, the values of P_m and $(dP/dt)_{max}$ are quite dependent on the particular material, particle size and shape, and moisture content for those dusts which can absorb moisture. Data on P_m and $(dP/dt)_{max}$ for very many combustible dusts are reported in a variety of references, with the most extensive compilations appearing in Field (1982) and NFPA68, 1988 Ed. Values for P_m are essentially independent of test vessel volume, but $(dP/dt)_{max}$ is quite dependent on this volume. The reported values may apply only for given size test vessels, or they may be scaled according to Bartknecht's cube-root law, which we discuss later. Some closed-vessel test data for selected combustible dusts are given in Appendix A.

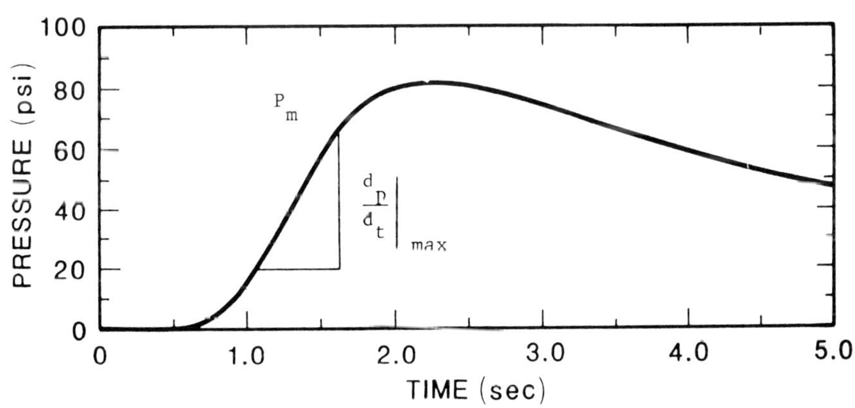

Fig. 2-1. Pressure in closed vessel test for a combustible dust.

In real industrial operations subject to potential dust explosion hazards, there is seldom, of course, the complete confinement provided in strong, closed vessel tests. Damaging explosions can and do occur in light metal enclosures such as cyclone separators or dust filter bag houses, piping and ducting for pneumatic conveyance of powders, various enclosures in grain elevators, etc. Few of these structures can withstand the full maximum pressures P_m for unvented dust deflagrations. Control and suppression methods are discussed later in this chapter.

Are detonations possible in any confined dust-air or dust-oxidizer mixtures? In long, strong tubes, detonations have indeed been induced for a variety of grain dusts, corn starch, fine metal powders, and coal dust [see Wolanski (1981)]. But, very strong ignition sources, usually detonation waves in explosive gas mixtures, are required to initiate such detonations.

An interesting exception is reported by Matsui & Kumamiya (1986). They report detonations in a strong, closed tube of fine soot films coated on the tube walls, initiated by both gaseous deflagrations and detonations. Slightly greater oxygen content than normally present in air was required in the tube atmosphere to achieve detonation. They also measured the effect on detonation pressures of increasing initial pressure up to 100kPa--one of the few studies which determined the effect of initial pressure greater than atmospheric. This study has considerable practical importance for accidents involving explosions in oxygen hoses.

Because it is seldom economical to test dust explosion effects and control measures full scale, one must know how to scale small-size test data to real industrial situations. The next section covers dust explosion scaling laws.

C. Scaling

It has been apparent for some time that the size and geometry of the confining vessel or structure has a strong effect on the development of the internal pressures during dust explosions. Similarly, the sizes and locations of vents, plus vent opening pressures and other parameters affecting venting of such explosions, are equally important in control measures to limit the maximum explosion pressures.

Because it is often too expensive or dangerous to measure or test dust explosion effects or control methods at full scale, small-scale testing using various types of instrumented enclosures has long been used instead. But, questions may then arise concerning the adequacy of small-scale test results and their extrapolation to full scale.

The first investigators who apparently addressed problems in scaling of dust explosion pressures in any depth were Donat and Bartknecht [see Bartknecht (1978a) and (1980)]. They determined, apparently in a simple empirical manner by testing in vessels of different volumes, that the maximum pressure rise in a closed vessel caused by an internal explosion was essentially independent of the volume, for a given combustible dust and dust-air suspension. They also found that the maximum rate of pressure rise was inversely proportional to the cube root of the vessel volume. This latter scaling law was expressed as Bartknecht's "Cubic Law",

$$(\dot{p})_{max} \cdot V^{1/3} = K_{st} \tag{1}$$

Bartknecht [1978a) and (1981)] also has developed and presents a series of nomographs for sizing vents to limit dust explosion pressures to some desired safe level. [These nomographs now also appear in NFPA (1988).] The reasoning Bartknecht presents for developing these nomographs is consistent with scaling vent areas according to:

$$\overline{A} = (A_v / V^{2/3}) \tag{2}$$

where $\overline{A}$ is chosen to limit maximum pressure to some desired level.

In trying to obtain a more fundamental understanding of development and venting of dust explosion pressures, we decided a few years ago to conduct a thorough similitude analysis of this problem [Baker, et al (1978)]. We were interested not only in the dust explosion itself, but also the ignition source, the internal and structural parameters of the enclosure, and ambient conditions and responses inside and outside the enclosure. Using principles of similitude theory, a series of non-dimensional ratios were derived. These ratios, or Pi terms, constitute a quite general scaling or model law for vented dust explosions.

1. Model analysis

A model analysis is performed by listing all of the parameters relevant to the problem, and applying the principles of similitude analysis [Baker, et al (1973)] to derive the non-dimensional ratios. The appropriate list for the problem of vented dust explosions is given in Table 2-1. From these 49 parameters, the 45 non-dimensional ratios listed in Table 2-2 can be derived. This is a formidable number of parameters to combine in an analysis. Fortunately, an experienced analyst can identify these parameters which are either invariant or those which are essentially constant over the range of interest, and therefore can be eliminated from the analysis without severe loss of accuracy. For example, the heat of combustion, H_d, does not vary greatly for combustible dusts which can explode. Additionally, this parameter appears in the Pi terms either to the first power or to the one-half power, which indicates that the heat of combustion should have a small effect on the venting of dust explosions. A complete discussion of the logic for selecting the Pi terms which significantly contribute to the venting of dust explosions appears in Hokanson and Seals (1978).

TABLE 2-1 Pertinent Parameters for Dust Explosions in a Vented Chamber

Parameters	Symbol	Dimensions
<u>Dust</u>		
Mass	M_d	M
Heat of combustion	H_d	L^2/T^2
Mean particle size	d_d	L
Particle size distribution function	f_d	-
Moisture content (mass ratio)	R_w	-
Heat of vaporization of moisture (internal energy)	U_{fg}	L^2/T^2
Dust concentration	R_d	M/L^3
<u>Ignition Source</u>		
Energy	E_i	ML^2/T^2
Duration	t_i	T
Volume	V_i	L^3
Geometry	ℓ_{ij}	L
Temperature	θ_i	θ
<u>Enclosure (Internal)</u>		
Volume	V_e	L^3
Vent area	A_e	L^2
Geometry	ℓ_{ei}	L
Initial pressure	p_{oe}	M/T^2L
Initial temperature	θ_{oe}	θ
Ratio of specific heats	γ_e	-
Total internal heat generation	q_{gen}	M/LT^3
<u>Enclosure Structure</u>		
Density	ρ_s	M/L^3
Elastic modulus	E_s	M/T^2L
Poisson's ratio	ν_s	-
Yield stress	σ_{ys}	M/T^2L
Other constitutive properties	σ_{si}	M/T^2L
Strains	ε_{si}	-
Displacements	δ_{si}	L
Thermal (radiative) absorptivity	α_{si}	-
Thermal conductivity	k_s	$ML/T^3\theta$
Initial temperature	θ_s	θ
Stefan-Boltzmann constant	σ	$M/T^3\theta^4$
Volumetric heat capacity	$\rho_s c_p$	$M/LT^2\theta$
Emissivity	ε	-

TABLE 2-1 (Continued)

Parameters	Symbol	Dimensions
Diaphragm burst pressure	p_b	M/T^2L
Mass per unit area of vent cover	m_v	M/L^2
<u>Ambient Conditions (Outside)</u>		
Pressure	p_∞	M/T^2L
Temperature	θ_∞	θ
Ratio of specific heats	γ_o	-
Relative humidity	h_o	-
<u>Response or Output (Internal)</u>		
Internal pressure	p	M/T^2L
Rate of pressure change	$dp/dt = \dot{p}$	M/T^3L
<u>Response or Output (Internal)</u>		
Loading times	t_ℓ	T
Temperature	θ	θ
Burning velocity	S	L/T
Flow velocity through vent	u_v	L/T
Mass flow through vent	Q	M/T
<u>External Response</u>		
Distance	R	L
Blast wave overpressure	P_s	M/T^2L
Blast wave times	t_b	T
Blast wave impulse	I_s	M/TL

TABLE 2-2. List of Pi Terms for Dust Explosions in a Vented Chamber

NUMBER	TERM	NUMBER	TERM	NUMBER	TERM
Dust		Enclosure Structure		Ambient Conditions	
π_1	$M_d H_d / p_{oe} V_e$	π_{16}	$\rho_s H_d / p_{oe}$	π_{31}	p_{∞} / p_{oe}
π_2	$d_d / V_e^{1/3}$	π_{17}	E_s / p_{oe}	π_{32}	$\theta_{\infty} / \theta_{oe}$
π_3	f_d	π_{18}	γ_s	π_{33}	γ_o
π_4	R_w	π_{19}	σ_{ys} / p_{oe}	π_{34}	h_o
π_5	U_{fg} / H_d	π_{20}	σ_{si} / p_{oe}	Response or Output	
π_6	$R_d U_d / p_{oe}$	π_{21}	ε_{si}	π_{35}	p / p_{oe}
Ignition Source		π_{22}	$\delta_{si} / V_e^{1/3}$	π_{36}	$\dot{p} V_e^{1/3} / p_{oe} H_d^{1/2}$
π_7	$E_i / p_{oe} V_e$	π_{23}	α_{si}	π_{37}	$t_L H_d^{1/2} / V_e^{1/3}$
π_8	$t_i U_d^{1/2} / V_e^{1/3}$	π_{24}	$K_s \theta_{oe} / p_{oe} H_d^{1/2} V_e^{1/3}$	π_{38}	θ / θ_{oe}
π_9	V_i / V_e	π_{25}	θ_s / θ_{oe}	π_{39}	$S / H_d^{1/2}$
π_{10}	$\ell_{ii} / V_e^{1/3}$	π_{26}	$\sigma \theta_{oe}^4 / p_{oe} H_d^{1/2}$	π_{40}	$U_V / H_d^{1/2}$
π_{11}	θ_i / θ_{oe}	π_{27}	$\rho_s c_p \theta_{oe} / p_{oe}$	π_{41}	$Q H_d^{1/2} / p_{oe} V_e^{2/3}$
Enclosure (Internal)		π_{28}	ε	External Response	
π_{12}	$A_e / V_e^{2/3}$	π_{29}	p_b / p_{oe}	π_{42}	$R / V_e^{1/3}$
π_{13}	$\ell_{ei} / V_e^{1/3}$	π_{30}	$m_v H_d / p_{oe} V_e^{1/3}$	π_{43}	P_s / p_{oe}
π_{14}	γ_e			π_{44}	$t_b H_d^{1/2} / V_e^{1/3}$
π_{15}	$q_{gen} V_e^{1/3} / p_{oe} U_d^{1/2}$			π_{45}	$I H_d^{1/2} / p_{oe} V_e^{1/3}$

2. Correlations with existing data and analysis

If we look at terms π_{12}, π_{35}, π_{36}, and π_{37} from Table 2-2, they say that, all other things being equal, pressures in the scaled enclosures should be unchanged; pressure rates should scale inversely as the cube root of the chamber volume; and all times should scale in proportion to the cube root of the chamber volume [see Eq. (1)].

The quantity K_{st} has dimensions of

$$K_{st} = \frac{ML}{T^3L} = \frac{M}{T^3} = \frac{FT^2}{LT^3} = F/LT \tag{3}$$

This is a direct consequence of invariance of term π_{36}, and Bartknecht's statement of equivalence of maximum pressure is a consequence of term π_{35}. Bartknecht's design nomograms are based on this "cubic law" and also utilize term π_{12} directly, as well as π_{35} and π_{36}.

An alternate form for Bartknecht's nomograms can be used, highlighting the importance of the dimensionless vent area ratio, π_{12}, in determining maximum pressure, π_{35}. A typical double set of nomogram curves yield only the three curves of Fig. 2-2, when plotted in dimensionless format using the Pi terms.

The correlation shown here is the only one we have found which can be compared to results of the model analysis.

3. Discussion of the scaling law

This model analysis can be used as a guide to future systematic studies, either experimental or analytical, relating to various aspects of dust explosion phenomenology. One should probably consider the Pi terms in groups in guiding the study, to determine the relative importance of various parameters. Then, the recommended procedure is to:

- o hypothesize a functional format for the model,
- o construct a test matrix,
- o conduct tests, and
- o incorporate the test results into the model.

Based on the results of the model analysis, the functional format for the empirical model is in general terms:

$$\begin{pmatrix} \text{Responses} \\ \text{or} \\ \text{Output} \end{pmatrix} = f \begin{bmatrix} \text{dust characteristics} \\ \text{ambient conditions (inside)} \\ \text{ignition source} \\ \text{ambient conditions (outside)} \\ \text{structure of enclosure} \end{bmatrix} \tag{4}$$

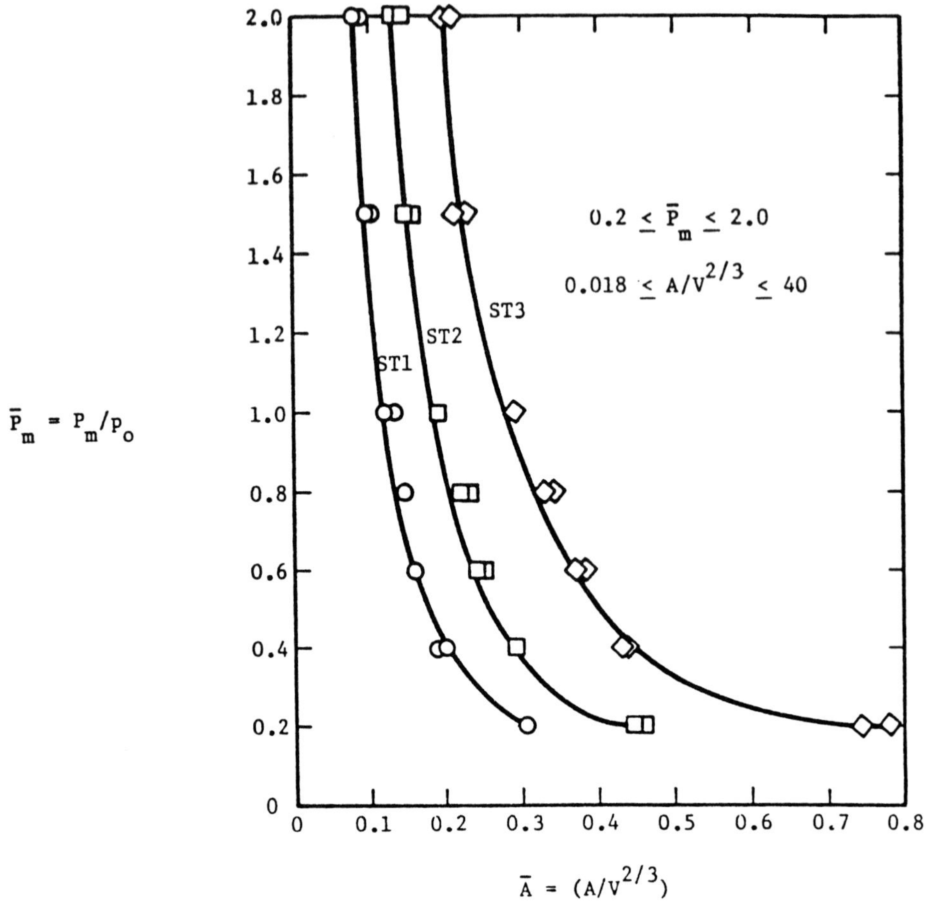

Fig. 2-2. Scaled Maximum pressures for dust explosions vs. scaled vent area ratio [Baker, et al (1978)].

This general format includes the effects of venting. The development of the ignition model implied by Equation 4 is a very formidable undertaking. To reduce the model development to a reasonable undertaking, several restrictions can be placed on the model. The model could be developed for dust initiation in a closed structure only, with no venting. Secondly, the initial temperature of the vessel could be constant and only one type of dust considered. With these restrictions, the functional format of the model reduces to:

$$\begin{pmatrix} \text{Responses} \\ \text{or} \\ \text{Output} \end{pmatrix} = f \begin{bmatrix} \text{dust characteristics: concentration} \\ \text{particle size, moisture content} \\ \text{environment: humidity} \\ \text{ignition source: type, intensity} \\ \text{duration} \end{bmatrix} \tag{5}$$

The objective of the model development would be to determine what conditions are necessary to initiate a dust cloud. This means that in each test, the primary measurement would be whether or not the dust cloud explodes. Additional parameters which should be measured in each test are the maximum pressure developed, maximum rate of pressure rise and the duration of the pressure (loading time). Writing Equation 5 in terms of the model analysis results in:

$$\begin{bmatrix} \text{Probability of} \\ \text{Ignition} \\ pV^{1/3}/P_{oe}H^{1/2} \\ p/p_{oe} \\ t_{\ell}H_d^{1/2}/V_e^{1/3} \end{bmatrix} = f\left(\frac{R_d H_d}{p_{oe}}, \frac{d_d}{v_e^{1/3}}, R_w, H_{oe}, \frac{E_i\, p_{oe}}{V_e}, \frac{t_i\, H_d^{1/2}}{V_e^{1/3}} \right) \tag{6}$$

The procedure to derive experimentally the interrelationships between the various Pi terms in Equation 6 and the probability of ignition is: a) develop the relationship between the probability of ignition and any one Pi term while holding the others constant, b) combine the individually derived relationships into an overall model, and c) conduct confirmatory tests to validate the model. As an example, suppose we want to derive the relationship between moisture content and the probability of ignition. The experimental procedure would be to select the dust concentration, particle size range, air humidity and ignition source characteristics. Then a series of tests would be performed in which the moisture content of the dust was varied over a range wide enough to establish the 100% and 0% probability of ignition limits. The moisture content would have to be varied over a range of about 1% to 14%. This series of tests would result in a curve similar to the one in Figure 2-3.

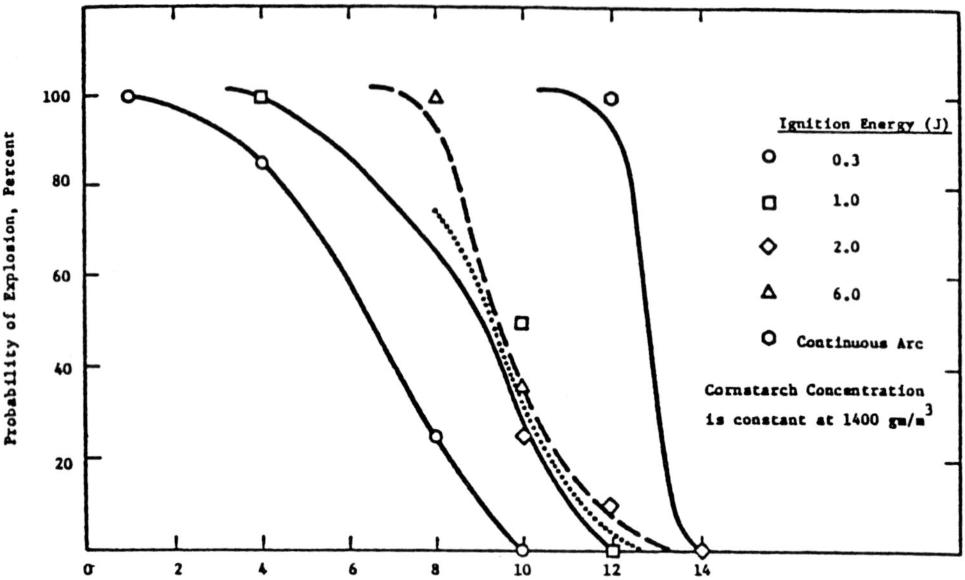

Fig. 2-3. The effect of moisture content on probability of ignition as a function of spark energy [adapted from Dirkzwager (1977)].

Similar investigations would be performed for each of the other Pi terms in Equation 6. Once all six parameters have been investigated, functional forms expressing the probability of ignition P(I), as a function of each Pi term (all other Pi terms constant), Equation 7, would be developed.

$$P(I) = f_1\left(\frac{R_d H_d}{p_{oe}}\right)$$

$$P(I) = f_2\left(\frac{d_d}{V_e^{1/3}}\right)$$

$$P(I) = f_3(R_w)$$

$$P(I) = f_4(H_{oe})$$

$$P(I) = f_5\left(\frac{E_i P_{oe}}{V_e}\right)$$

$$P(I) = f_6\left(\frac{t_i H_d^{1/2}}{V_e^{1/3}}\right) \qquad (7)$$

The next stage of model development would be to combine in some fashion each of the six limited relationships into a general empirical relation. There exists no set procedure for deriving such a general relation; the only basis is the experience and background of the analyst. To date, the author has seen no attempts at such correlations for combined Pi terms.

D. Testing methods

Nearly all evaluations of dust explosions must rely on test results. The designs, sizes and uses of explosion test vessels or other apparati vary considerably. We give here a rather brief review of the apparati and methods most commonly used. Eckhoff (1987) discusses a number of small-scale test equipments, and we paraphrase some of his descriptions.

1. Hartmann apparatus

The first apparatus developed to run controlled tests for minimum ignition energy of dust clouds was apparently the Hartmann apparatus, designed by I. J. Hartmann of the U.S. Bureau of Mines. It is shown schematically in Fig. 2-4.

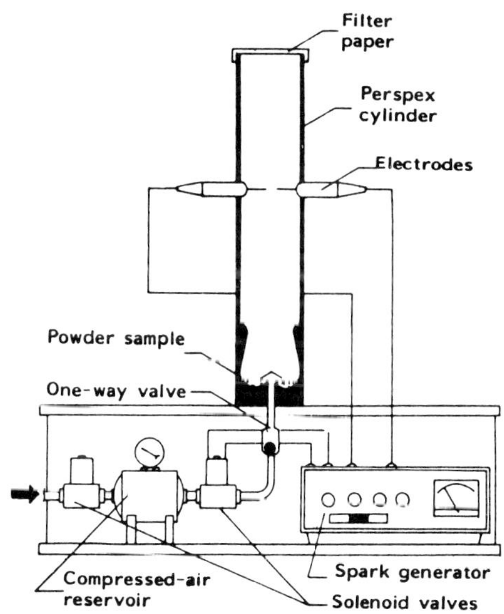

Fig. 2-4. Apparatus for establishing the minimum ignition energy of a dust cloud (The Hartmann Apparatus) [Eckhoff (1987)].

A dust sample is placed in the bottom of the 1.2 liter cylindrical chamber, a measured volume of compressed air is blown into the chamber, providing a transient dust-air mixture. Usually, a continuous spark ignition source tests ignitability. Ignition is determined usually by flame and by rupture of the filter paper diaphragm at the top of the tube. This apparatus has been used for many years to measure ignitability limits for a wide variety of dusts; at first by the U.S. Bureau of Mines and later by many other investigators in the U.S. and abroad.

A spark generator designed at Christian Michaelson Institute (CMI) produces sparks of any energy between 4.5 mJ and 2.9 J. Spark energies are calculated from recorded traces of spark voltage and current, which are integrated over the discharge duration.

Before testing begins, the optimum combination of amount of powder, dispersion air pressure, and delay between dust dispersion and spark discharge, is determined in initial trials. Ten explosion trials are then performed at each of a series of successively increasing or decreasing spark energies to determine the frequency of ignition at each energy. Ignition is defined as such when a flame fills the tube and the filter paper bursts. Frequency of ignition versus energy is plotted on a graph. E_{min} (the minimum ignition energy) is defined to lie between the highest energy-value that did not produce any ignitions, and the lowest energy-value that produced at least one ignition in 10 trials.

2. Hartmann pressure bomb

Hartmann also designed a small pressure vessel test chamber, with the same 1.2 liter volume as the apparatus for determining minimum ignition energy. A modern version of the Hartmann Bomb is shown in Fig. 2-5.

In the version used at CMI, the ignition source is an electric spark of approximately 3.5 J net energy and has a discharge time of about 2-3 ms. Compressed air from a 60 cm^3 reservoir at 6-8 bar above atmospheric is used for dispersing the dust. A range of dust concentrations are tested, in agreement with common practice. Two parameters are extracted from each recorded pressure-versus-time trace, namely the maximum slope of the trace during the build-up of pressure $(dP/dt)_{max}$ and the peak pressure P_{max}. In general, results obtained in series of replicate tests scatter quite substantially. Comprehensive investigations at CMI have disclosed that the results can be assumed to be distributed normally. Test results are therefore presented in terms of arithmetic mean values, standard deviations and expected 95% levels, $(m+1.65\,\sigma)$. The highest 95% value is taken as the test result.

In Section IIC, we discussed scaling of the maximum pressure rise rate $(dP/dt)_{max}$, and presented Bartknecht's "Cubic Law" for such scaling as Equation 1. Bartknecht (1978a) showed by testing in the Hartmann Bomb and in larger closed test vessels up to 1 m^3 volume that the Hartmann Bomb gave unconservative results for scaled values of this parameter for many dusts. Eckhoff (1984) discusses this topic in more detail, and points out that the cubic law must be applied with caution; and should only apply for situations with similar dust

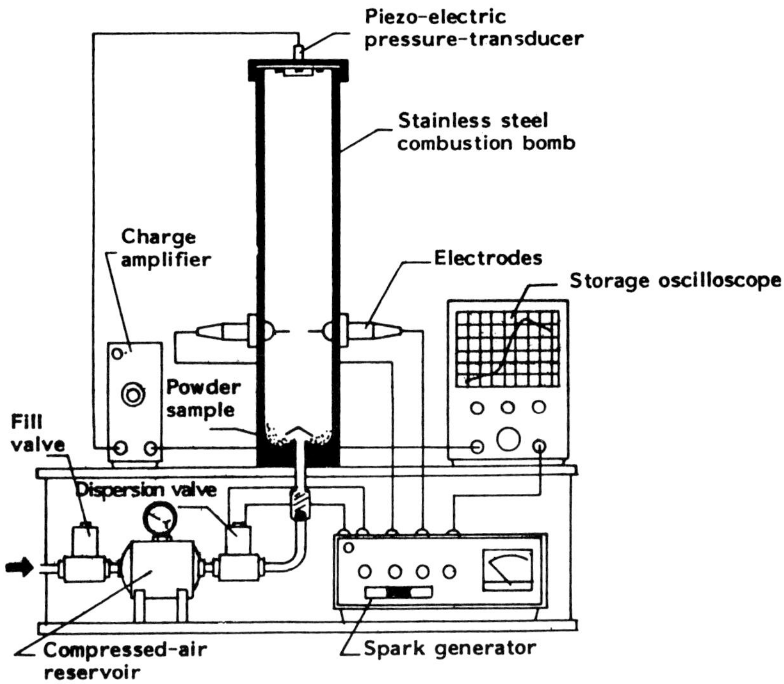

Fig. 2-5. Small-scale apparatus for measuring P_{max} and $(dP/dt)_{max}$ (The Hartmann Bomb) [Eckhoff (1987)].

dispersion turbulence, ignition source properties, and etc. From our model analysis, we certainly concur with Eckhoff's opinion.

Regardless, the work of Bartknecht and others has led to development of wide use of larger closed vessel apparati of various volumes, which will be discussed next.

3. Intermediate-scale vessels

Bartknecht (1978a) developed a series of roughly spherical test vessels of several volumes. In his testing, he concluded that a vessel of about 20 liters volume was required to avoid underestimating $(dP/dt)_{max}$ via Equation 1. Several designs of vessels with this volume have evolved and are now in wide use. Two such designs are shown in Figs. 2-6 and 2-7. The Bartknecht vessel is spherical, and employs a split, perforated ring for dust dispersion. The U.S. Bureau of Mines vessel is of different geometry, and employs a bottom dispersion nozzle. Sapko, et al (1987) report good correlation with test data for coal dust explosions from the U.S. Bureau of Mines 20 liter chamber and measurements taken in the Bruceton Experimental Mine.

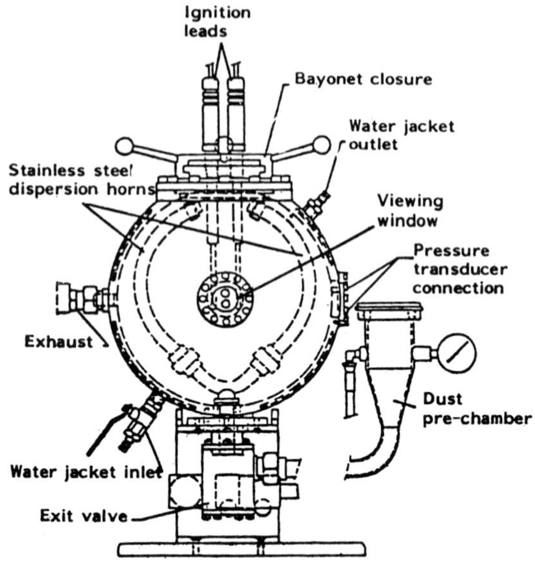

Fig. 2-6. Bartknecht 20 liter spherical explosion bomb [Eckhoff (1987)].

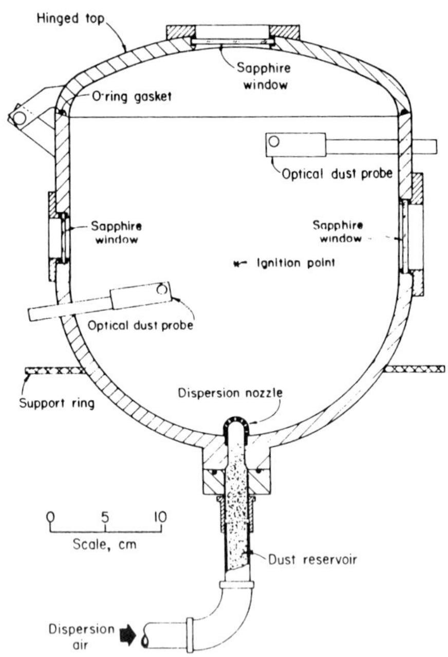

Fig. 2-7. U.S. Bureau of Mines 20 liter laboratory explosibility test chamber [NFPA68 (1988)].

In operating the Bartknecht 20 liter apparatus, dust is first placed in the pre-chamber (see Fig. 2-8) and the pressure inside this chamber is then raised to 20 bar(g). Following the opening of the exit valve, the dust sample is blown into the bomb through small holes in the dust dispersion horns, forming a dust cloud. After a pre-determined delay, an ignition source is activated at the center of the sphere. The standard source is a 10 kJ pyrotechnical ignitor. The chamber is evacuated to 0.4 bar (abs.) before commencing the dispersion/ignition sequence. This is in order to achieve atmospheric pressure inside the bomb following dust dispersion and just prior to ignition.

Resulting explosion pressures and rates of pressure rise are measured using two pressure transducers fitted into the side of a vessel. When desired, other initial pressures than atmospheric can easily be generated by adjusting the bomb pressure prior to dust dispersion. The water jacket also makes it possible to work with other initial temperatures besides the ambient one.

Enright (1984) reports testing in an approximately 8-liter chamber (actually, 7.8 liters), which he states was based on a U.S. Bureau of Mines design. This chamber is shown schematically in Fig. 2-8. The base cup and dust dispersion system appear to be identical with the base cup in a Hartmann Bomb.

Enright (1984) shows in closed-vessel tests of volumes 1.2 through 20 liters with lycopodium, cornstarch and aluminum powder using a continuous spark ignition source that scaled maximum pressure rise rates, K_{st}, generally <u>increased</u> with vessel size. This is in general agreement with Bartknecht's earlier results.

4. <u>Cubic meter vessels</u>

Bartknecht's work has lead to construction and use of a number of closed test vessels with volume of about one cubic meter. Other laboratories have built similar test vessels. One such vessel is shown in Fig. 2-9. Dust dispersion is obtained in this vessel with a perforated ring dispersion system (Fig. 2-10), somewhat similar to that illustrated in the 20-liter sphere in Fig. 2-6. NFPA68 (1988) discusses problems of clogging of this dispersion system, and recommends that a whipping hose injector be used instead.

Field (1982), as noted earlier, includes extensive tables for closed-vessel tests in the Hartmann Bomb, but also includes considerable data from tests in cubic meter vessels. Selected data from cubic meter vessel tests are included in Appendix A of this book, taken from NFPA68 (1988).

Senecal (1988) reports testing of dust explosion suppression systems in both 190 liter (0.190 m^3) spherical vessels, but more extensive testing in a 1.9 m^3 spherical vessel. The larger vessel is shown schematically in Fig. 2-11. He also reports using pyrotechnic squibs of known energy release as the ignition source. This is a relatively common practice in dust explosion testing in larger vessels.

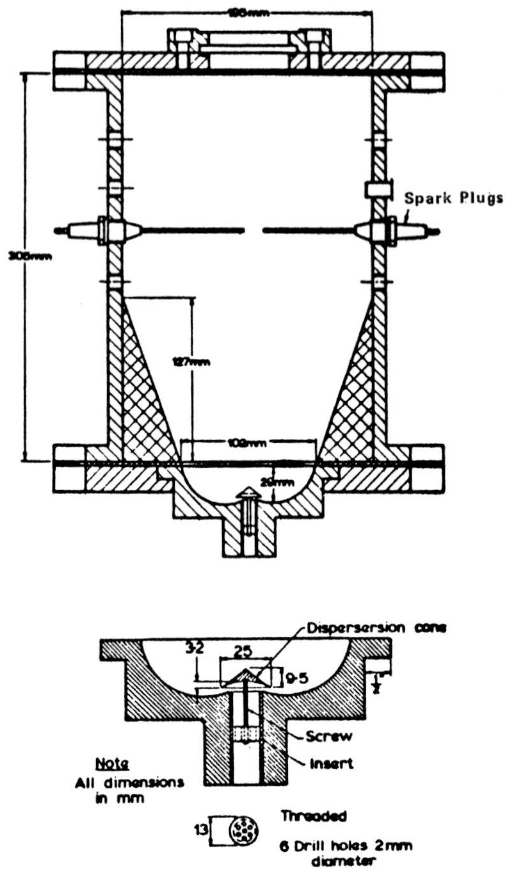

Fig. 2-8. Eight liter explosion chamber.

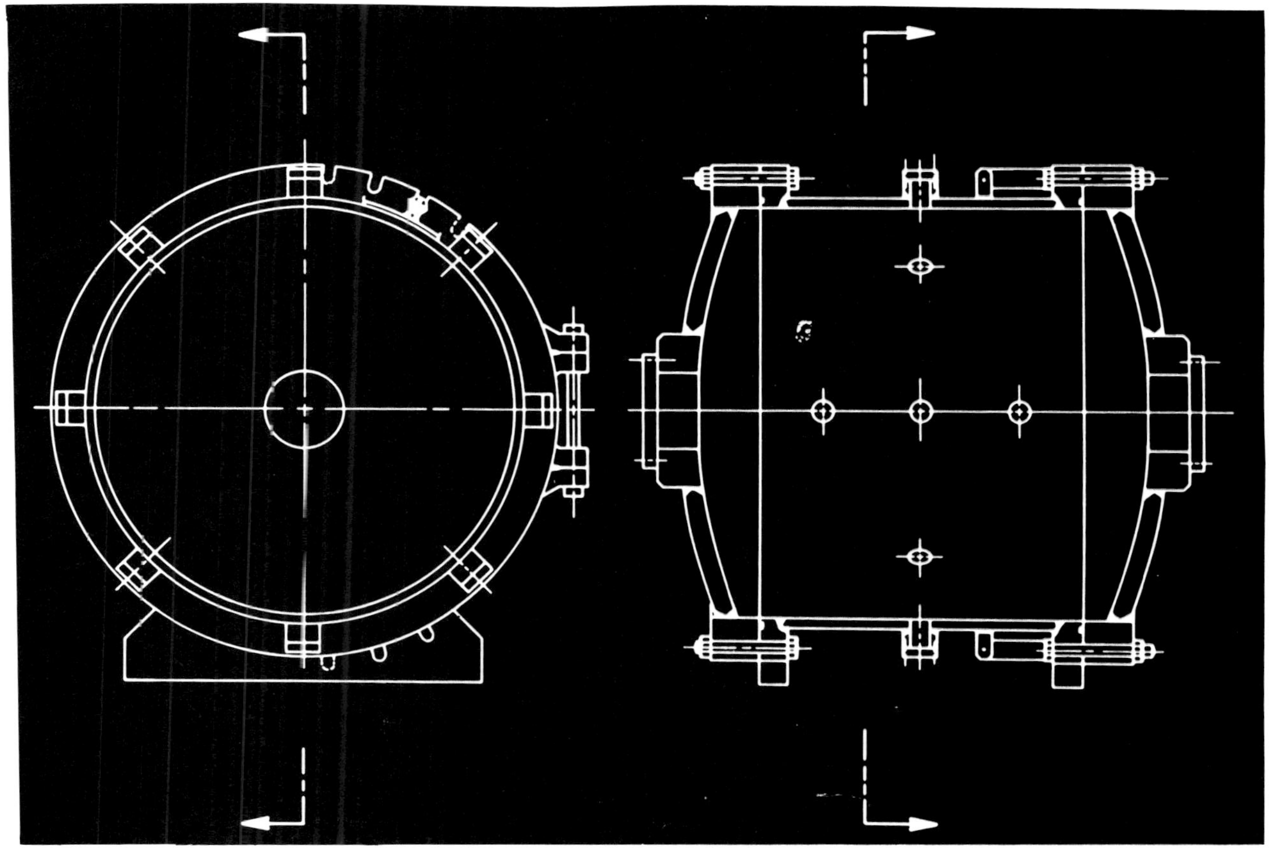

Fig. 2-9. Southwest Research Institute cubic meter vessel for dust explosion investigations. (Courtesy of Southwest Research Institute)

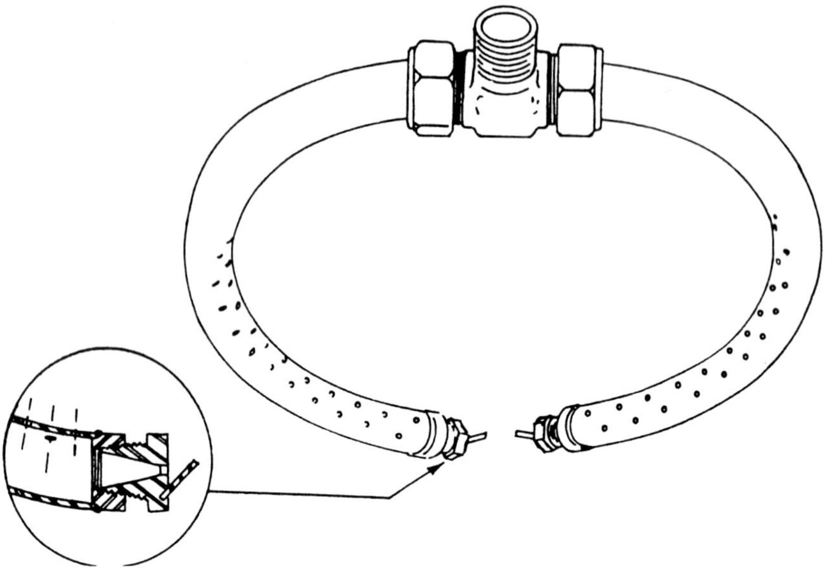

Fig. 2-10. Perforated ring dispersion system for cubic meter vessel [NFPA68 (1988)].

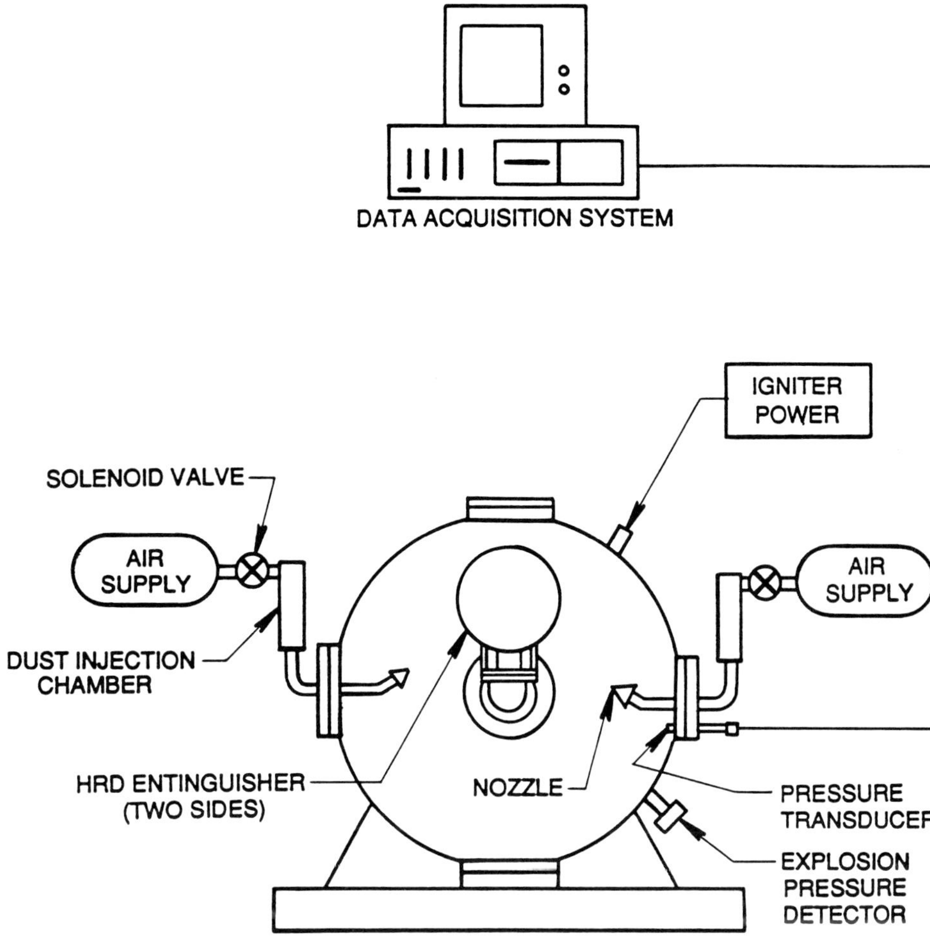

Fig. 2-11. 1.9 cubic meter vessel for suppression testing.

5. Large scale testing

We have already mentioned testing of coal dust explosions in the Bruceton Experimental Mine as an example of large-scale dust explosion testing (in this case, full scale). But, other large-scale test vessels or other apparati have been used for special purpose testing. Some examples follow.

Bartknecht (1986) describes a 250 m^3 spherical test vessel for evaluating effectiveness of explosion suppression and venting systems, and for verifying scaling. He notes that test data for contained and vented explosions scale well from small scale tests.

Eckhoff (1988) reports testing in three models of grain silos (tall cylindrical vessels) over a range of volumes from 20 m^3 up to 500 m^3. A schematic of a 236 m^3 steel silo in Norway is shown in Fig. 2-12. Length-to-diameter ratios for all three silo models are nearly 6:1. Eckhoff (1988) notes

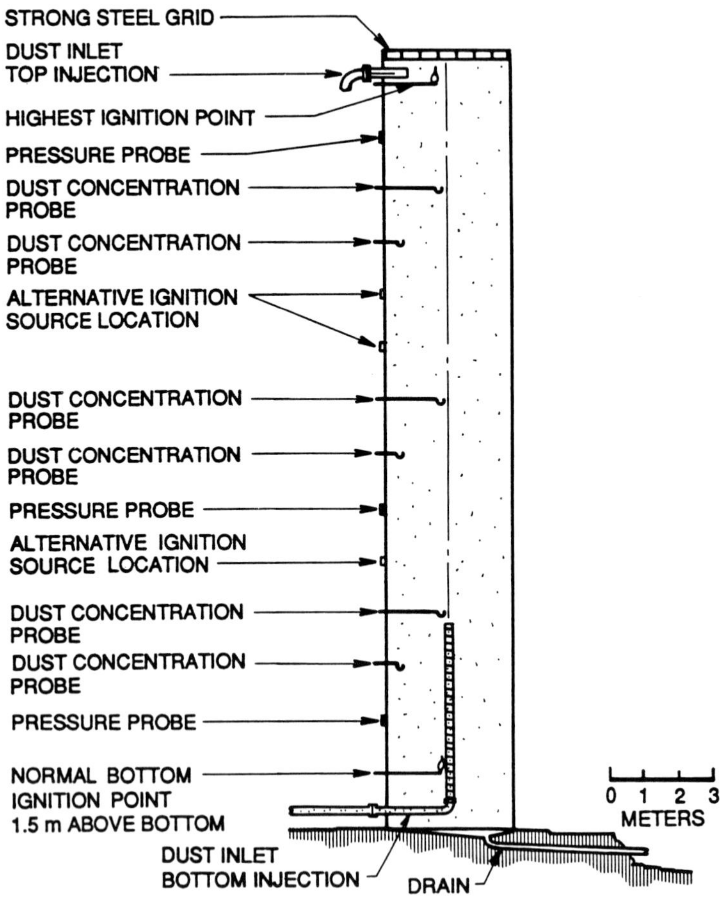

Fig. 2-12. Experimental cylindrical 236 cubic meter steel silo at Sotra, outside Bergen, Norway.

that maxima for vented pressures measured during tests in these silos are quite dependent on method of dust injection, as well as ignition source location. (Please note that this does not prove a failure of scaling, but instead shows dependence on more exact replication of all important scaled parameters.

Our final example of dust explosion testing in a relatively large chamber is reported by Pineau (1984). He describes testing in a vented 100 m^3 horizontal chamber dug out of a limestone cliff, as shown in Fig. 2-13. The testing is related to control of explosions in milk powder plants.

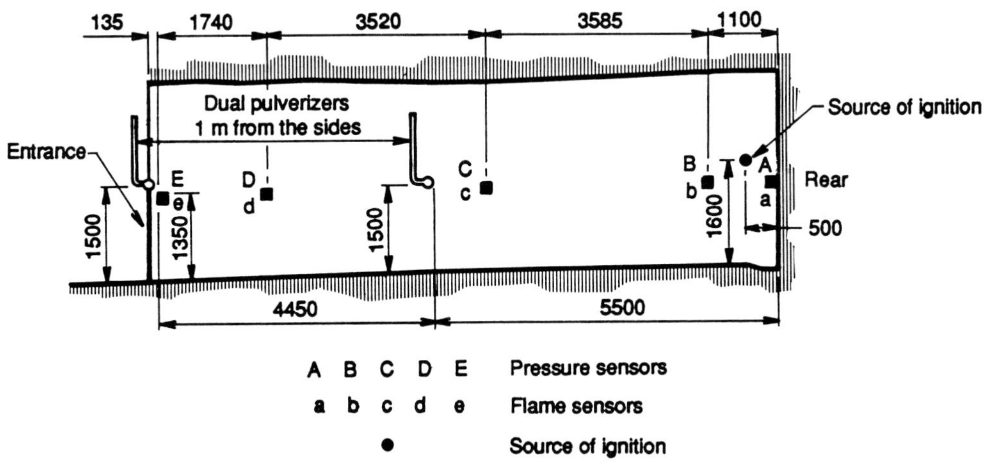

Fig. 2-13. 100 cubic meter explosion chamber.

6. Other special apparati

In his excellent review paper, Eckhoff (1987) discusses many types of tests and apparati for determining ignitability and explosibility of dust clouds. We have already discussed much of the test equipment he reviews, but other special apparati noted in his paper and by Field (1982) are discussed here.

In a test for dust ignition adopted by U.S. Bureau of Mines [Dorsett, et al (1960)], a basket of metal mesh is filled with the powder and placed in a furnace through which air of constant, known temperature is flowing at a slow, specified rate. The temperature within the powder bed is monitored continuously, and by increasing the air temperature in steps, a level is reached at which the temperature in the powder sample begins to rise above that of the surrounding air. This critical air temperature is taken as the spontaneous ignition temperature of the powder in question. This temperature is not a true powder constant but depend on the experimental conditions. In particular the critical temperature will depend on the amount of powder tested.

In the USA and the UK, the ignition temperature of dust clouds in contact with a hot surface has traditionally been determined in the Godbert-Greenwald furnace [Dorsett, et al (1960)].

This apparatus, which is illustrated in Fig. 2-14, consists of a furnace and a temperature-control unit. In the furnace, which is open at its lower end, a thermostatically-controlled heating element is wound around a ceramic lining.

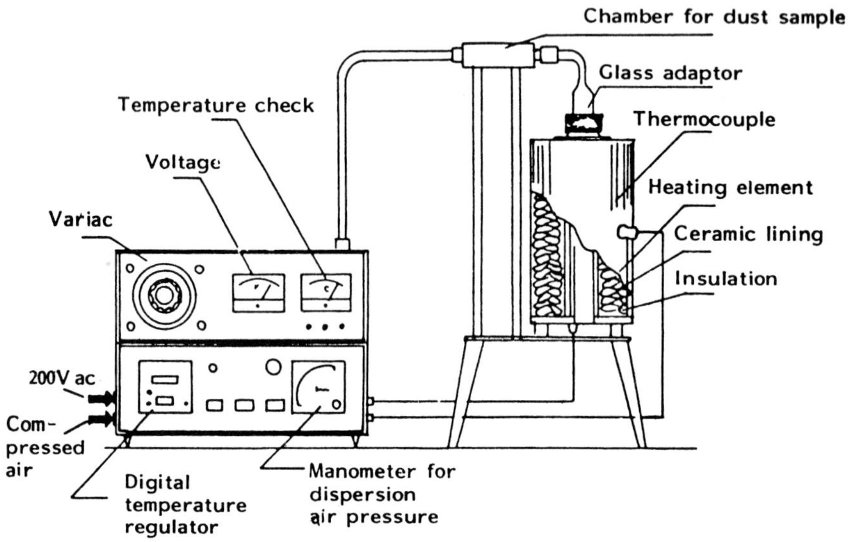

Fig. 2-14. Apparatus for measurement of minimum ignition temperature of dust clouds (The Godbert-Greenwald Furnace) [Eckhoff (1987)].

The upper end continues via a glass adaptor to a dust-dispersion chamber, which is in turn connected by a tube via a magnetic valve to a compressed air reservoir in the control unit. The temperature of the furnace is automatically regulated to a given level by an electronic temperature regulator. The powder is placed in the dispersion chamber and when the required temperature is reached, the compressed air is released from the reservoir and the dust is dispersed, passing through the heated part of the furnace in the form of a cloud. Ignition is observed in the form of a flame, emitting from the lower opening of the furnace. The temperature range of the apparatus is 50-1000°C.

One hot-plate test apparatus, which is shown in Fig. 2-15, consists of a modified electric hot-plate, a temperature-control unit, three thermocouple devices and a 2-channel plotter. The hot-plate may be held at a given temperature, which is read by one of the thermocouples and displayed on one of the plotter's channels. On the surface of the plate is laid a metal ring, with a diameter of 100 mm and a height of either 5 or 15 mm.

The powder sample to be tested is placed in the metal ring and carefully levelled off. A thermocouple is placed in the sample through a hole in the metal ring. The sample temperature is displayed on the second of the plotter's two channels. The third thermocouple acts as a control device for regulating the plate temperature.

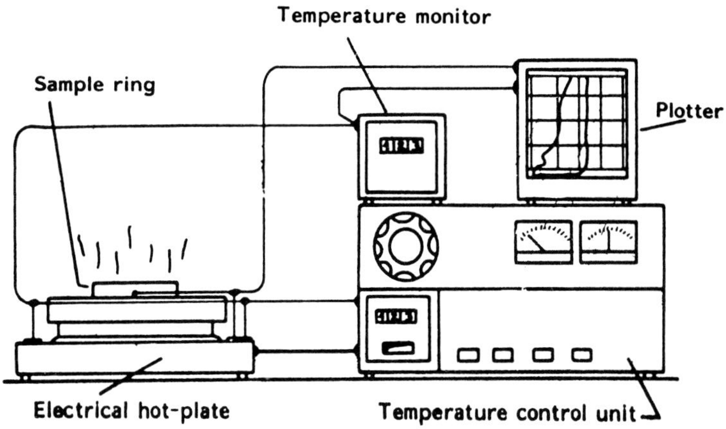

Fig. 2-15. Hot-plate test apparatus [Eckhoff (1987)].

The trials are allowed to continue until the sample has ignited (which is either observable or registered on the plotter). Alternatively, tests are terminated when the temperature in the sample has not increased beyond that of the plate.

The minimum ignition temperature is given by the lowest plate-temperature that gives ignition of the sample, rounded-off to the nearest temperature in °C that is divisible by 10. This value is for the chosen layer thickness. If a sample will not ignite using the maximum plate temperature, then the duration of the longest trial at that temperature, is reported.

Eckhoff (1987) reports that this apparatus is used by most of the leading European dust explosion laboratories.

Another apparatus for determining flame ignition of dusts is shown in Fig. 2-16. The main purpose of tests with this apparatus is to determine whether or not dust explosions using a given powder in air, at normal pressure, are at all likely. The tests is therefore qualitative, although some quantitative information is also obtained. A vertical steel tube of length 40 cm and diameter 14 cm is mounted on a base-plate fitted with two holes into which a U-shaped dust dispersion tube and an acetylene welding torch nozzle are positioned. A quantity of the powder is placed at the bottom of the dispersion tube, and a blast from a compressed air reservoir disperses the dust into a cloud in the tube, which is then immediately exposed to the hot flame from the welding torch. The amount of powder and the dispersion air pressure are varied to produce the most severe conditions for ignition. When an explosion occurs, the maximum height of the flame, and the apparent violence of the explosion, are estimated by the observer.

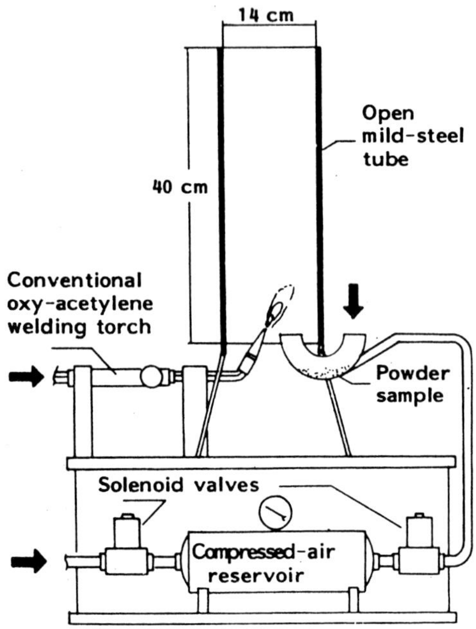

Fig. 2-16. Flame ignition apparatus [Eckhoff (1987)].

Eckhoff (1987) also describes a test apparatus used in Sweden for measuring minimum ignition energy of dust layers.

This apparatus (Fig. 2-17) is based upon a 2 mm-thick steel plate (a) in which 10 circular holes of 12 mm diameter have been made. This plate is positioned on top of the base-plate (b). The assembly of the two plates thus forms a number of circular cavities, into which the dust to be tested is poured and scraped off, leaving samples of 2 mm-thick powder layers ready for testing. A top-plate (c) covers all samples except the one to be tested. Electric sparks of various energies and durations are then passed through the sample from the upper needle electrode (e) to the earthed base-plate (d). The upper assembly (f-k) is adjusted so that the distance from the upper electrode is to the top surface of the powder layer is 1 mm. Twenty ignition trials are then performed at each of a series of successively increasing or decreasing spark energies to determine the frequency of ignition at each energy. Ignition is defined as such when the complete sample layer ignites and burns. Frequency of ignition versus energy is plotted on a graph. E_{min} (the minimum ignition energy) is defined to lie between the highest energy-value that did not produce any ignitions, and the lowest energy-value that produced at least one ignition in 20 trials.

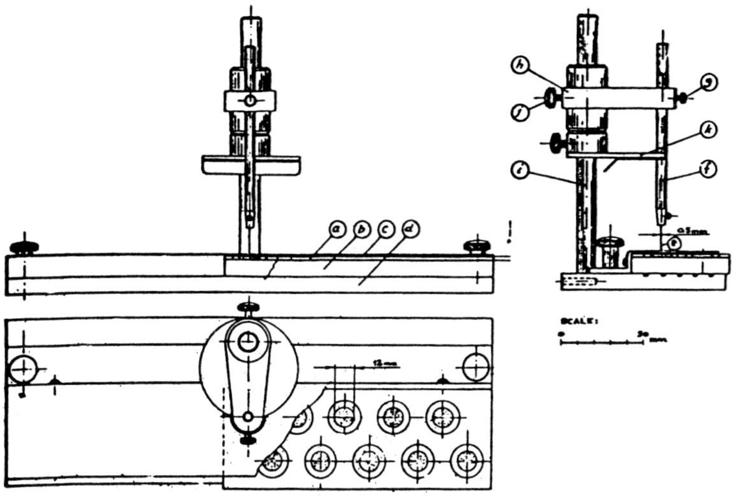

Fig. 2-17. The Nordtest apparatus for the measurement of minimum ignition energy of dust layers [Eckhoff (1987)].

In estimating or predicting maximum energy which can be released in a dust explosion, one should know the heat of combustion of the dust. Unfortunately, this property is almost never reported for combustible dusts in the dust explosion literature, but instead is given only in references on properties of fuels which can be burned in power plants or heating furnaces. Standard oxygen bomb calorimeters are readily available to make such measurements (Figs. 2-18 A&B). One places a carefully-weighed sample of the fuel in a cup in the oxygen bomb, with a fuse wire buried in the sample. The bomb is closed and sealed, and flooded with oxygen at several atmospheres pressure. It is then immersed in a water-cooled jacket. On ignition, the temperature rise of the known mass of water yields an accurate measure of heat of combustion of the sample.

At the University of Michigan and several other research centers in dust explosions, dust detonation tubes are used for much of their research. Fig. 2-19 shows a vertical detonation tube apparatus. Combustible dust is fed into the tube from the feed hopper. Detonation is initiated by igniting a detonable gas mixture in the initiator tube. Gas detonation bursts a diaphragm in the initiator tube, and the resulting strong shock initiations dust-air deflagrations or detonations in the vertical tube. Instrumentation includes pressure switches, pressure transducers and optical instrumentation such as Schlieren or shadowgraph. Lee, et al (1981) and Anonymous (1987) include a several papers giving test results with this apparatus.

Oxygen Bomb Calorimeter

A. Oxygen bomb

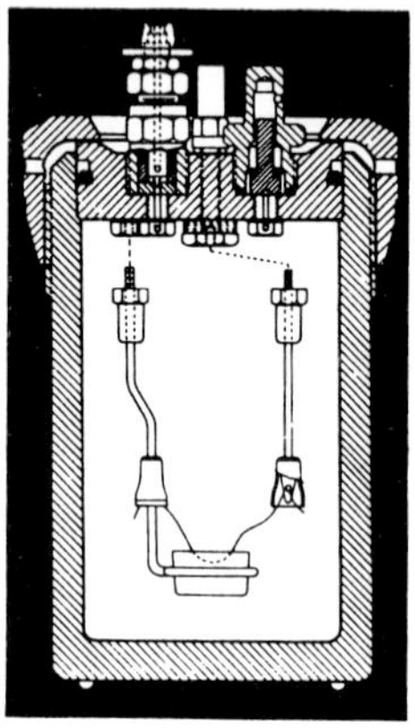

"Plain Oxygen Bomb Calorimeter" Bulletin No. 1300, July 1984; Parr Instrument Company, Moline, Illinois.

B. Adiabatic jacket

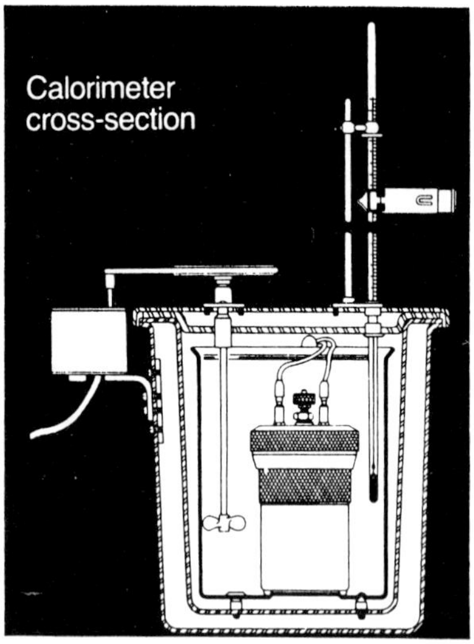

"Plain Oxygen Bomb Calorimeter" Bulletin No. 1300, July 1984; Parr Instrument Company, Moline, Illinois.

Fig. 2-18. Oxygen bomb calorimeter.

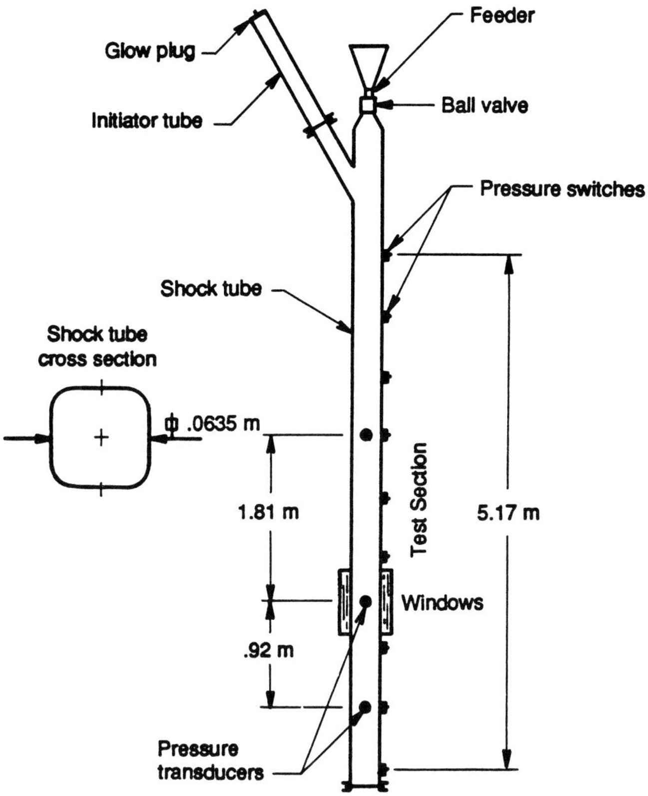

Fig. 2-19. Vertical dust detonation tube [Lee, et al (1981)].

7. Problems in testing

Dust explosions involve very complex physical phenomena. So, many of the testing methods we have described may not truly represent real dust explosions. This is particularly true for small-scale laboratory test apparati. Also, a recurring problem for confined dust explosion pressure testing is to assure uniform dust concentrations suspended in the apparatus at test time, or to measure the concentrations. We have illustrated dust dispersion devices in Fig. 2-4, 2-5, 2-6, 2-7, 2-8 and 2-10. These problems are discussed at some length in Eckhoff (1987) and NFPA68 (1988).

A common known or suspected cause of ignition in dust explosions in systems which transport dusts or powders pneumatically is discharge of static electricity. Special apparatus is available to observe or measure electrostatic charge buildup and discharge, but we do not discuss it in this book.

E. Control methods

Because accidental dust explosions can only occur within enclosures, and because the initial combustion and pressure rises during the explosions are relatively slow processes, various passive and active means of explosion control are possible. In this section, we review these methods.

1. Venting, small to intermediate scale

Passive venting of enclosures within which dust explosions can occur is a favored method of controlling internal deflagrative explosions and avoiding destroying the enclosures [Bartknecht (1978a), (1981) and NFPA68 (1988)]. Fig. 2-21 shows schematically the effect of venting with a covered vent designed to open at gage pressure P_v. Pressures can reach successive maxima, e.g., P_1, P_2, and P_3 in Fig. 2-20, but these maxima are all well below structural strength in a property designed vent system. Scaled vent area, $A_v/V^{2/3}$, when A_v is vent area and V is enclosure volume is an important parameter in controlling the maximum vented pressure. Bartknecht (1978a), (1981) has developed nomographs to aid in vent design with Fig. 2-21 being an example. We have also shown previously that the same predictions can be made from dimensionless curves based on similitude analysis (Fig. 2-2). These nomographs are also included in NFPA68 (1988), for three classes of dusts with different reactivities designated by ranges of K_{st} which are characteristic of dusts of different materials and average particle sizes. The three classes defined by Bartknecht are given in Table 2-3. NFPA68 (1988) also fits equations to the nomographs.

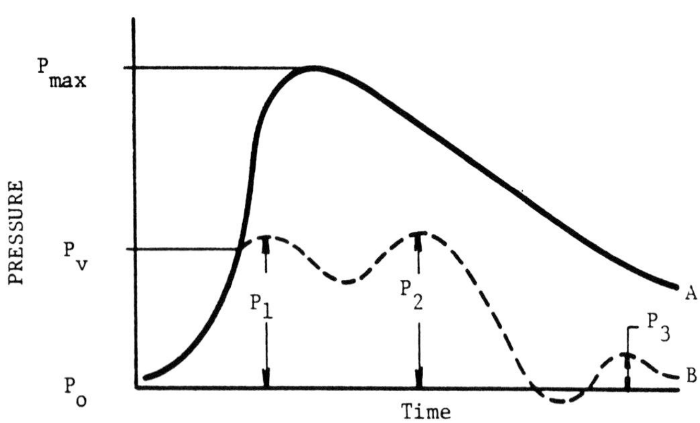

Curve A--unvented, Curve B--vented

Fig. 2-20. Time-pressure histories for unvented and vented deflagrative explosions [Baker, et al (1983)].

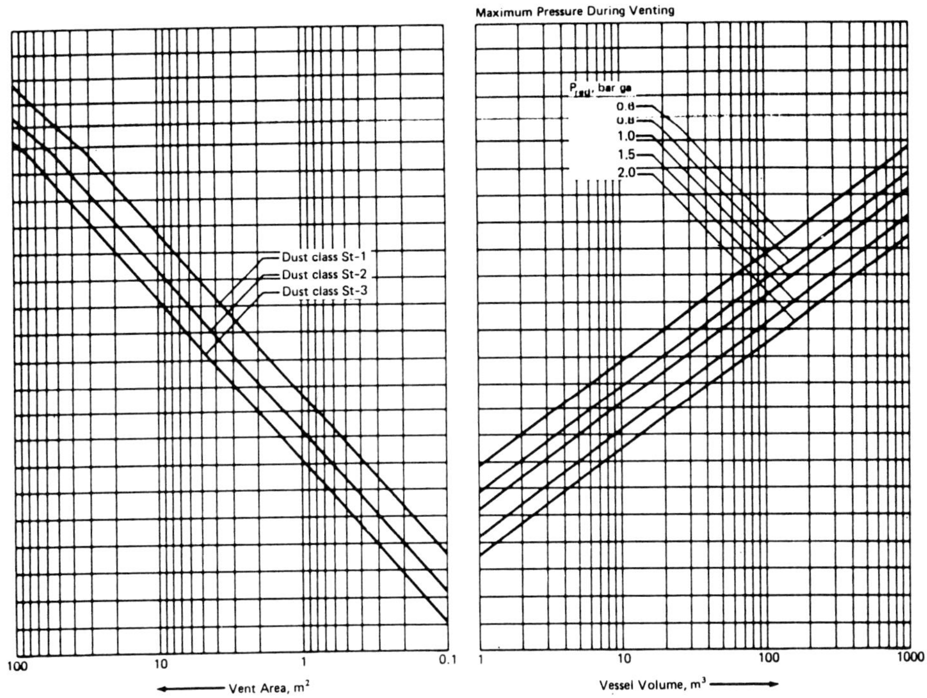

Fig. 2-21. Venting nomograph for classes of dusts--P_{stat} = 0.5 bar ga [NFPA68 (1988)].

TABLE 2-3 Bartknecht's Ranges of Values for Dust Reactivity, Optimum Dust-Air Mixtures.

Level	K_{st} range ($bar \cdot m \cdot s^{-1}$)
St 1	$1 < K_{st} < 200$
St 2	$201 < K_{st} < 300$
St 3	$301 < K_{st}$

Practically, if venting is to protect a structure or enclosure from internal dust explosions, one must vent either directly to the outside air or through short, large cross-section ducts. Venting through long ducts will restrict flow through the vent opening and increase pressure during venting to values much greater than P_{red}. Bartknecht (1978a), (1981) and NFPA68 (1988) discuss this constraint in some detail.

Nomograms such as Fig. 2-21 are also limited to dust explosion control by venting of enclosures whose principal linear dimensions are approximately equal (that is, typical vessels or room-shaped enclosures). Venting of long ducts or pipes must employ multiple vents at rather frequent intervals along the ducting.

2. Venting, large scale

In many industrial processes which can generate or store combustible dusts or dusty materials within enclosures, the enclosures can be very large, and they are also often nearly filled with the material being stored or processed. Common examples are grain silos, surge or supply bins in plastic plants, hoppers, etc. Surface area of the enclosures is often limited because most of the internal surface is often covered by the product in the enclosure. NFPA68 (1988) discusses this problem and practical aspects of installing venting, as well as operational constraints to assure that vents are never covered by the solid granular material in the enclosure.

Grain elevators are probably the largest structures susceptible to dust explosion damage. We will discuss such explosions and their control later in this chapter.

A secondary hazard associated with venting control is the formation of large fireballs outside the vent. This is mentioned in NFPA68 (1988), and is very evident in photos in Bartknecht's work [Bartknecht (1978a), (1981), (1986)]. But, no measurements seem to have been made of the thermal radiation intensity or duration for such fireballs, so this potential hazard is not well quantified. To the author, this seems to be a serious oversight in dust explosion hazards testing.

3. Inerting

Inerting of enclosures is another passive method of dust explosion control. In concept, this control method is simple. Combustible dusts cannot burn if no or little oxygen is present, so maintaining a constant inert atmosphere within a dusty enclosure will certainly prevent explosions. Typical inertant gases are nitrogen, carbon dioxide, steam, argon, helium, or flue gas. Synonyms for inerting used in various publications are blanketing, and oxidant concentration reduction.

Closed-vessel dust explosion test apparatus can be used to determine volumes of various inert diluent gases which will prevent burning of combustible dusts. Most such testing appears to have been done by the U.S. Bureau of Mines. Data on maximum oxygen concentration to prevent combustion of suspended dusts are reported in NFPA69 (1986), and are included in this book as Appendix B. This guide is the most widely accepted publication in the United States regarding inerting as an explosion control method.

Bartknecht also discusses inerting of dust explosions through vacuum and with solids.

One hazard sometimes ignored with inerting explosion control is the hazard of asphyxiation. Many reactive combustible dusts can only be inerted by reducing oxygen levels below life supporting levels. So one must take care in inerting areas which personnel must enter.

4. Deluge

As with most fires, combustion of dusts in buildings can be extinguished by water deluge if the combustion is detected before escalation to explosion occurs. So, sprinkler or other water deluge systems can control certain classes of dust explosions. Sensing of smoldering combustion can be by thermal or other detectors, triggering water deluge. Certain fast-acting deluge systems in closed vessels can be classed as active suppression systems.

5. Suppression

Dust explosion suppression as a control method implies an active system to sense early stages of the explosion and quench or mitigate the explosion by rapid flooding of enclosure with fire suppressant material. The method is briefly discussed in NFPA69 (1986), but is covered in much more detail by Bartknecht (1978a), (1981), Gillis (1987) and Senecal (1988).

Detectors which sense the initial stages of the explosion can be ultraviolet or infrared radiation detectors or sensitive pressure transducers. Testing in relatively large vessels has shown that active explosion suppression is possible for high reactivity (high K_{st}) dusts [Senecal (1988)] using several chemical suppressants. A typical deflagrative pressure measured in the 1.9 cubic meter test vessel of Fig. 2-11 is shown in Fig. 2-22, together with a trace for a suppressed deflagration. Bartknecht (1986) shows testing of suppression systems for large volume enclosures with the 250 cubic meter test vessel.

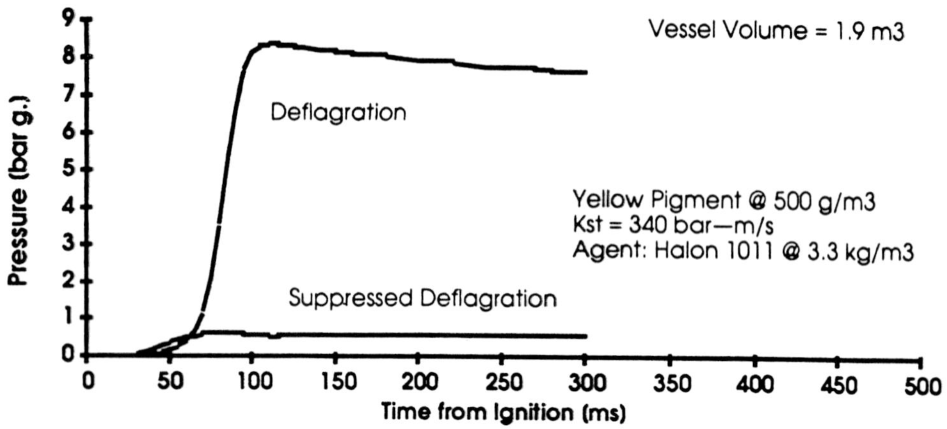

Fig. 2-22. Deflagration and suppression in a dust explosion.

F. <u>Dust explosions in grain elevators</u>

1. <u>Case histories</u>

With the exception of coal dust explosions in coal mines, the largest and most damaging class of dust explosions has been explosions in grain elevators. These elevators are designed to move large quantities of grain very quickly, transferring the grain from trucks and railcars to storage silos, and then to other bulk transport including ships. Generation of quantities of combustible dusts is inevitable in the rapid transfer of the grains in the elevators; there are many potential ignition sources; and a number of areas in the elevators have the partial confinement and complex geometry which can accelerate relatively benign small dust fires into very damaging explosions. These explosions have occurred in the earliest grain elevators, and they continue to occur to the present. They are probably the most complex of all explosions, and the most difficult to assess and control or mitigate.

Kaufman (1986) gives a good history of grain elevator dust explosions, noting the first report of a dust explosion in a flour mill in Turin, Italy, in 1785. He gives case histories of several explosions since 1979 in the United States. His paper also incudes summaries of fourteen grain elevator explosions between January 1979 and April 1981, which caused totals of 20 fatalities, 58 injuries and property loss exceeding 37 million dollars. Kaufmann (1986) notes that the U.S. Department of Agriculture has kept track of agricultural dust explosions since 1980, and he gives an extensive table listing these reported explosions. In the seven-year period covered in Kaufmann's paper, 154 accidents were reported resulting in 49 deaths and 224 injuries.

It is apparent that there is a continuing dust explosion problem in the agricultural industry.

2. The National Academy of Sciences study

From the previous section, it is apparent that damaging dust explosions are rather frequent occurrences in grain elevators. But, in December 1977, five such explosions occurred in the United States within a week, resulting in 59 deaths, 48 injuries, and drastic reduction of our grain export capability. In July 1978, at the request of the U.S. Department of Agriculture, the National Academy of Sciences conducted an international symposium on grain elevator explosions. Shortly after the symposium, the U.S. Occupational Safety and Health Administration asked the National Academy of Sciences to form a ten-member study group on this topic, and such a panel was formed in November 1978. Objectives were:

1. Study the federal government's investigation of grain elevator explosions and make recommendations for improvement,
2. Investigate grain elevator explosions selected by OSHA occurring during the panel's tenure and determine their causes,
3. Recommend actions to be taken to reduce the occurrence of explosions in grain elevators and mills and to identify any needed research and development,
4. Produce a handbook on dust collection methods and systems for grain elevators and mills, and
5. Develop a methodology for investigating explosions in grain-handling facilities based on the panel's experience.

The panels' approach was based on risk management methodology. The steps were:

- o Definition of situation or location where risks are to be controlled, i.e., dust explosions in grain elevators.
- o Identification of causes of grain dust explosions via hazards scenarios.
- o Evaluation of identified risk. This was done by ranking identified hazards into a hierarchy of importance, using three measures--severity of hazard, frequency or probability, and resources to control the hazard.
- o Control of significant risk. The panel sought expert counsel and visited selected grain-handling facilities to address this topic.
- o Financing of uncontrolled risk. The panel discussed the options of loss write-off, assumption of debt, self-insurance and other means of transferring the risk or loss.

To illustrate the complexity of the geometry and possible sources of dust explosions in grain elevators, we include a section through a typical elevator with sources for dust clouds identified as Fig. 2-23. Part of the process of

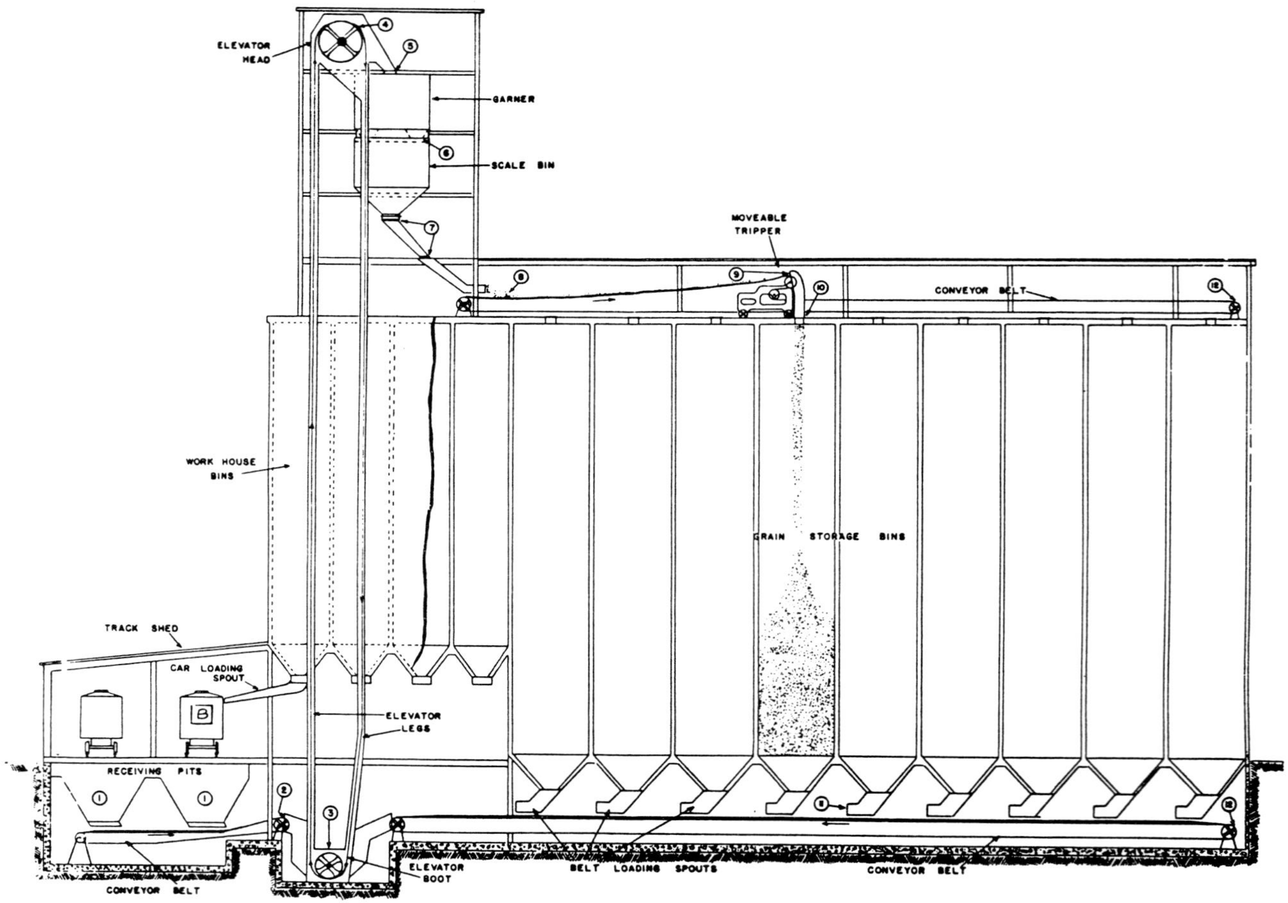

Fig. 2-23. Diagrammatic section view of a terminal type grain elevator. Circled numbers indicate points at which dust clouds are likely to be emitted [Anon. (1980)].

identifying dust explosion hazards scenarios employed by the panel was to use Fault Tree Analysis. The resulting fault tree is shown in Fig. 2-24. Fig. 2-25 gives the results of the third step in the panel's evaluation, evaluation of risk.

Conclusions of the study were:

1. Numerous reports on explosion prevention in the grain-handling industry have been published over the past 60 years. Many present long lists of actions to be taken with little or no assessment of their feasibility or potential effectiveness or explanation of why they were needed. These reports generally do not consider the influence of human characteristics on explosion prevention, questions requiring research, or methods for disseminating widely the available information on explosions and their prevention. In addition, the existence of these reports appears to be little known.
2. Recent compilations would make it appear that the number of grain-handling facility explosions occurring annually in the United States has increased during the past 20 to 25 years but that much of this apparent increase could be due to better reporting procedures. Nonetheless, it appears that the problem of grain elevator explosions is greater than generally realized.
3. The elevator leg is the most dangerous location with respect to initial of the primary dust explosions.
4. Grain dust is generated in many places in elevators and mills. Of particular concern is the dust in confined spaces that, without proper housekeeping, will accumulate in layers on all surfaces and present a potential for secondary dust explosions.
5. There is a considerable body of documented evidence indicating that electrostatic discharge can ignite dust clouds under the right conditions; however, the panel found no evidence of ignition due to electrostatic discharge in its investigations of explosions.
6. The contribution of human operatives and external factors other than the immediate physical aspects (e.g., people's attitudes, insurance practices, and government regulation) often are a major part of the problem and often are overlooked.

The panel recommended a number of actions by the grain-handling industry and by government, or by cooperative efforts. They divided these actions into three priorities. The panel believed that the first-priority actions should be implemented in <u>all</u> facilities and that the second- and third-priority actions should be implemented to the extent possible depending on the specific facility. There was no internal ranking within each category. Specific recommendations appear in Appendix C.

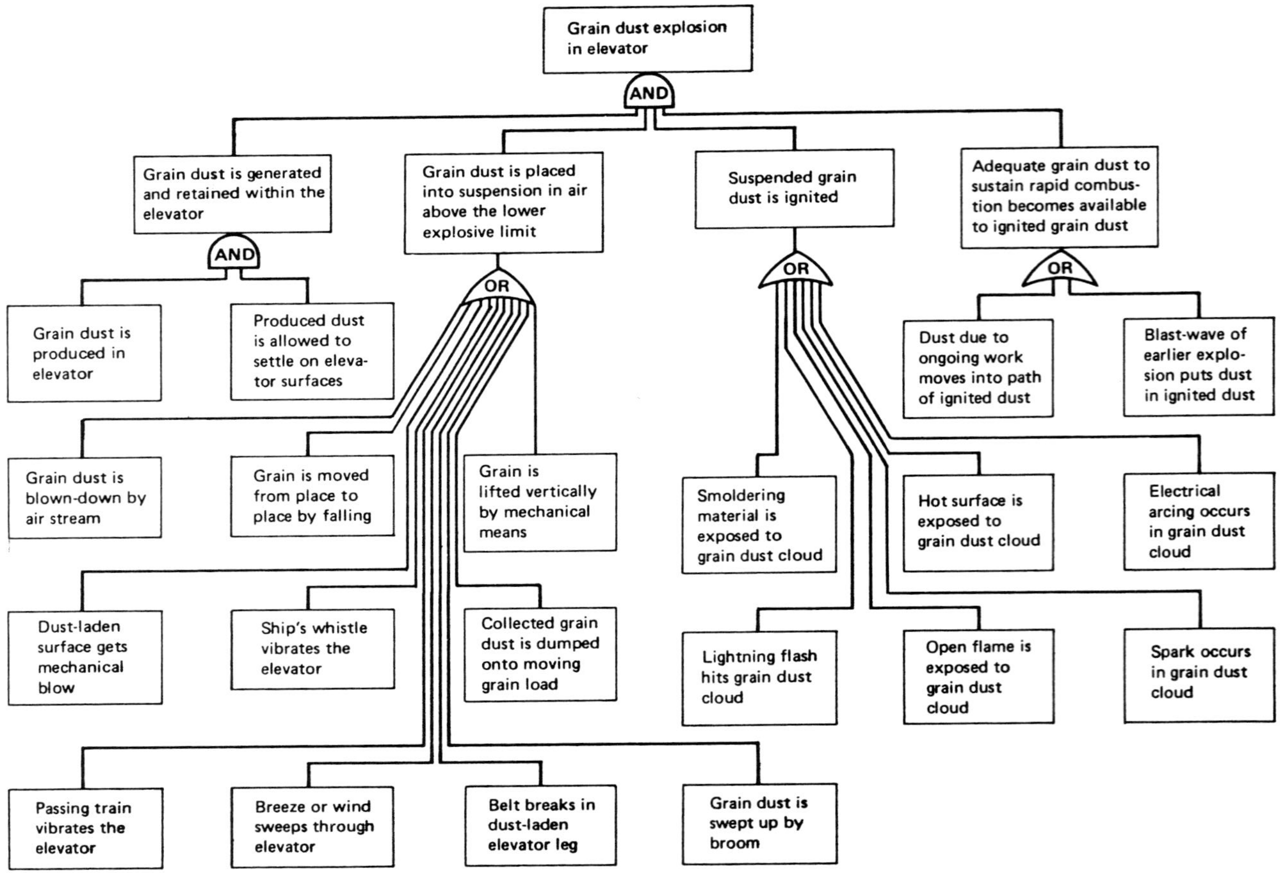

Fig. 2-24. Fault tree analysis for grain elevator explosions [Anon. (1980)].

Hazard Severity

Code	Hazard Severity	Effect on Elevator Objectives	Effect on Functional Capability	Effect on Personnel Safety
A	Catastrophic	Total loss of product with no salvage. All customer services terminated	Physical plant is destroyed or damaged beyond effective use.	Elevator employees, bystanders, or others are killed.
B	Critical	Majority of product lost. Only partial salavge possible. Customer services reduced to low level.	Two or more functions are disabled; elevator must be shut down.	Major injuries occur to employees or bystanders.
C	Marginal	Only 75% of product is salvagable. Grain services are possible by improvisation.	Temporary disruption of elevator functions; normal operations can be restored in 1 day.	Minor injuries occur to employees or bystanders
D	Negligible	No significant effect on product or service to customers.	No apparent damage to elevator operation.	No apparent harm to employees or others.

Hazard Probability

Code	Description of Situation
J	The identified hazard scenario could occur on the average of once a week.
K	The identified hazard scenario could occur on the average of once a month.
L	The identified hazard scenario could occur on the average of once a year.
M	The identified hazard scenario could occur on the average of once a decade.

Hazard Elimination Control Resources

Code	Calculated Dollar Equivalence (Value of all resources required to either eliminate or control the identified hazard scenario; revision of policy, procedures, manpower, dollars, technology, facilities, materials, and schedule.)
P	Preventive action for the identified hazard will require less than $2,500.
Q	Preventive action for the identified hazard will require between $2,500 and $50,000.
R	Preventive action for the identified hazard will require between $50,000 and $250,000.
S	Preventive action for the identified hazard will require more than $250,000.

Fig. 2-25. Hazard scenario ranking criteria for grain elevator explosions [Anon. (1980)].

G. Closure

We hope that this chapter will introduce the reader to the complex topic of dust explosions, and provide enough references to the voluminous literature on this topic to at least serve as a guide to that literature. Much work on this topic is continuing worldwide, so the interested reader is strongly urged to follow the current international literature to stay abreast of new developments.

One serious hazard from dust explosions has not, in the author's opinion, been given enough attention in research on this topic. That is characterizing the fireballs from these explosions, and their thermal radiation effects. These fireballs can be very large, and radiate very significant thermal energy, as is quite apparent from color photos of fireballs for vented dust explosions [Bartknecht (1978a), Cashdollar & Hertzberg (1986)]. Yet, we could find no measurements of the fireballs or their effects.

III. REACTIVE GAS EXPLOSIONS

A. Introduction

Explosions of reactive gases have occurred and been investigated for more than a century. However, they have again received considerable attention in recent years in connection with the safety aspects of large scale production, transport, and storage of liquified petroleum and natural gases. The safe operation of off-shore platforms that handle petroleum and natural gases is of particular concern. The Three Mile Island incident in 1979 also raised the question of safety in nuclear powerplants relating to hydrogen-air explosions in a nuclear reactor containment. Two-phase hybrid explosions are also of concern in accidental fuel release in chemical and petrochemical plants and in coal mines and powder industry. The two-phase explosions are similar to gas explosions but they involve more physical processes prior to combustion reactions. They will be discussed in more detail in Chapter IV.

The aim of this chapter is to provide a relatively comprehensive understanding of the physical phenomena and the prediction of the damage potential from reactive gas explosions. Emphasis will be given to summarizing experimental results carried out during recent years, particularly large-scale field tests. Systematical descriptions of combustion and detonation theories will be excluded, since they have been well presented in many books, e.g., Williams (1985) and Fickett and Davis (1979). Nevertheless, basic concepts and illustrative conclusions of some theoretical analyses are included to keep this chapter self-contained.

1. Terminology: explosion, deflagration, detonation

Before proceeding with the topic of reactive gas explosions there will be a brief review of terminologies associated with subjects to be covered later. It will be helpful since established usage of certain words related to explosion phenomena can be misleading.

Explosion: Explosion is a term which corresponds to rapid heat release, energy release, or pressure rise. Explosions can, of course, occur in non-reactive, pressurized gas vessel failures (see Chapter V). However, an explosive gas is a medium which will permit rapid energy release by chemical reactions, and hence usually refers to certain premixed gases of fuel and oxidizer.

Furthermore, gas explosions do not require the transmission of a wave through the explosion source medium, although explosions in most scenarios involve some kinds of waves. In other words, an explosion does not necessarily require the passage of either a deflagration or a detonation wave through the exploding medium. A volumetric explosion can be taken as an example of explosions without wave propagation. In volumetric explosions, an explosive mixture contained in a vessel is suddenly heated to a high enough temperature for rapid reactions to occur simultaneously.

To the contrary, both deflagrations and detonations are waveforms propagating through explosive gases. The differences between deflagration and detonation will be described extensively later and some brief mention here will clarify the synonyms and similarities.

Deflagration: A deflagration is a subsonic wave sustained by chemical reactions. In general, a combustion wave is considered as a deflagration only, although the word combustion is very diverse in nature.

It is general practice to call a combustion wave or a deflagration wave a flame or burn. Therefore in the normal sense, deflagration, combustion, flame, and burn are synonymous and have been used interchangeably [Glassman (1977)].

Detonation: According to Fickett and Davis (1979), a detonation is a shock wave sustained by chemical reactions. The leading part of a detonation is a strong shock wave propagating into the explosive medium. This shock heats the medium by compressing it, thus triggering chemical reactions, and a balance is attained such that the chemical reaction supports the shock.

The term detonation has been uniquely used so far to refer to the rapid and violent form. Thus, no further clarification is needed in the brief review.

2. Differences between deflagration and detonation

Deflagration and detonation are both waveforms propagating in reactive media and both are supported by the energy release from chemical reactions in the media. However, they differ from one another substantially in many ways, as described in the following.

A detonation propagates at supersonic speed relative to the unreacted gas ahead of the wave, with typical velocities of the order of a few kilometers per second, whereas a deflagration is a subsonic wave with typical velocities of the order of meters per second. In other words, a detonation propagates a thousand times faster than a deflagration.

As the consequence of the high propagation velocity of the detonation wave, which is much faster than the speed of pressure equalization, there will be significant pressure differences across the wave front. In other words, significant overpressure will be generated upon the arrival of a detonation wave, even in open air. The pressure at the detonation front is about 15 to 20 times the initial pressure for most hydrocarbon fuels mixed with air, and the value will be doubled for that mixed with pure oxygen. On the contrary, the pressure difference across a flame front is so small that the slow mode of combustion can be treated as a constant pressure process. This is because the equalization of pressure occurs at acoustic speed, which is much faster than the flame propagation. As a consequence, the overpressure generated by a deflagration in an open atmosphere is negligible. It should be noticed that this is not to say that the slow burn proceeding in an enclosure will not cause the overall pressure level inside the enclosure to rise. In fact, an overall pressure rise of 6 to 8 times the initial pressure will result due to the slow burn of most hydrocarbon-air mixtures in an enclosure [see Baker et al (1983), Chapter 1]. This pressure rise is sufficient to destroy the confinement of many enclosures, such as fuel tanks, ship compartments and buildings. However, if the flame remains laminar with low subsonic speed, the pressure rise throughout the confinement is spatially uniform.

On the other hand, the pressure rise in detonation process is highly localized at the wave front and the pressure field is highly directional and asymmetric. Nevertheless, the overall pressure generated by a detonation in the enclosure after equalization remains the same as that generated by a slow burn, providing that the confinement is not ruptured during the process.

In essence a deflagration may be considered as an expansion wave where pressure and density drops and gas velocity increases across the wave. Conversely, a detonation is a compression wave and relative to the wave the flow velocity decreases while pressure and density increase. Furthermore, for a supersonic detonation wave, the unburned gas ahead of the wave front is undisturbed and remains in its initial state; whereas a deflagration is a subsonic wave and can thus perturb the gas ahead of the flame. As a consequence, there exists a unique detonation velocity (as well as other detonation parameters) for the given mixture with specified initial conditions. On the contrary, the deflagration speed is relative to the moving unburned mixture in front of the flame and hence cannot be uniquely determined by the given initial conditions.

The most important difference between a deflagration and a detonation is perhaps the different propagation mechanisms. The diffusion of heat and species from the reaction zone to the unburned gases is responsible for the initiation of chemical reactions ahead of a deflagration whose speed is thus limited by the molecular diffusivities to the order of a meter per second. On the other hand, adiabatic shock compression of the unburned gases is responsible for the initiation of chemical reactions in a detonation wave. In this process, the flame burns in highly compressed and preheated gases and burns with extreme rapidity; thus, material is consumed thousands of times faster than in a normal flame. Typical propagation velocities of the order of kilometers per second result.

A deflagration can be ignited by a weak energy source such as an electric spark or a match head with an energy of only a few millijoules. However, the instantaneous initiation of a detonation requires a rather powerful energy source with an energy several orders of magnitude higher; i.e., at least higher than joules. Nevertheless, strong ignition sources are not essential to detonation, since abrupt change from deflagration to detonation can occur under appropriate conditions.

3. Turbulent flame propagation

An explosive gas may, in general, support either a deflagration or a detonation wave depending upon various conditions, the most obvious being mixture compositions, ignition source, and confinement. Deflagration and detonation are two extreme modes of combustion with characteristics distinguished as discussed above. In between the two extremes there is an almost continuous spectrum of burning rates ranging from the laminar flame speed to supersonic velocities in excess of the Chapman-Jouguet detonation speed. A laminar deflagration can readily be triggered by a weak ignition source such as an electric spark or a match head. However, it is unstable by nature because the expansion of combustion products generates compression waves in unburned gas and causes the flame to accelerate. The positive feedback and other mechanisms of flame acceleration will bring the flame speed to as high as hundreds of meters per second or even to some critical point for detonation to occur.

It is interesting that in both laminar deflagration and detonation processes, turbulence plays a negligible role; whereas in cases between the two extremes, turbulence dominates. In accident scenarios, the in-between cases are most frequently encountered since a laminar flame is unstable and the onset of detonation requires more restrictive conditions. Turbulent flame propagation is very important from the standpoint of safety engineering, not only because of its frequent occurrence in accidental explosions, but also due to the violence level of high speed turbulent flame which is not much less than that of a detonation.

4. Organization of chapter

This chapter is organized somewhat differently than other chapters. That is partly because this is the longest chapter, reactive gas explosions being of primary interest for accidental explosion assessment, and as a result triggering extensive analyses and experimentation worldwide. Also, it seemed more logical in the presentation of this material to discuss each critical subtopic and issue related to reactive gas explosions rather completely, interrelating theory, experiment, and experimental apparatus as we proceeded. Further, Prof. M. J. Tang of East China Institute of Technology wrote much of this material, and the writing reflects her extensive experience on this topic, as well as her style of writing.

B. Scaling

In this chapter, we discuss several classes and a wide range of rates of reaction for reactive gas explosions. The dominant physical processes can be quite different for some of the different explosion classes, and also quite different over different reaction rate regimes. So, it is appropriate to develop or present separate scaling laws for certain restricted cases.

1. Scaling of vented gas deflagrations

Bartknecht's "Cubic Law" [Bartknecht (1978a) and (1981)] is essentially a scaling law for vented gas deflagrations. It can be stated as:

$$P_m = f_1\,(\text{stoichiometry, type of combustible gas, } A_v / V^{2/3})$$

$$(\dot{p})_{max} \times V^{1/3} = K_g \quad (1)$$

Also, Bradley and Mitcheson's (1978a and b) analyses and data comparisons for pressures in vented explosions of this kind imply a scaling law, with:

$$\frac{P_m}{p_o} = f_2\,(\overline{A}, \overline{S}_u) \quad (2)$$

where

$$\overline{A} = A_v K_d / A_s \quad (3)$$

and

$$\overline{S}_u = S_u / a_o \quad (4)$$

In these expressions,

P_m = maximum explosion (gage) pressure
P_o = ambient pressure
$(\dot{p})_{max}$ = maximum rate of pressure rise
V = vessel volume
K_g = a "constant" dependent on type of gas and mixture stoichiometry
A_v = vent area
K_d = discharge coefficient
A_s = interior surface area for assumed spherical vessel

S_u = gas velocity ahead of flame front

a_o = unburned gas acoustic velocity

In many tests with approximately spherical vented vessels up to a cubic meter in volume, all data scaled well with Equations (1) or (2). But Zalosh (1980) then ran a test series on larger, room-size vented enclosures of about 35 m^3 volume, and found late time, oscillatory pressure peaks which far exceeded the initial pressure peaks. This work triggered other testing to this and larger scales, [see Eckhoff, et al (1980) and Van Wingerden and Zeeuwen (1983)], which showed that these results were reproducible, and that the amplitudes of the late, oscillatory pressures <u>increased</u> with the scale of the testing. Pappas, et al (1983) have proposed empirical methods for estimating this phenomenon.

In any event, it seems certain that a <u>simple</u> scaling law for vented gas explosions is inadequate, and a more complex law must be stated. Such a law is presented here, with parameters chosen on the basis of the experimental results already noted, and on known or postulated physical parameters which could affect the venting process, including the late time instability. Other work reviewed in developing the law appears in Oyez (1984), Lee, et al (1982), Sapko, et al (1976), Chippett (1980), Lee and Guirao (1982b), Nagy, et al (1971), and Yao (1974).

A simplified schematic for a vented chamber containing a combustible gas mixture which is ignited by some ignition source, is shown in Fig. 3-1. Only a few of the related physical quantities are indicated in this figure. We assume that, in addition to these more obvious parameters, a number of others could be important in the physics of burning, vent opening, and combustion instabilities.

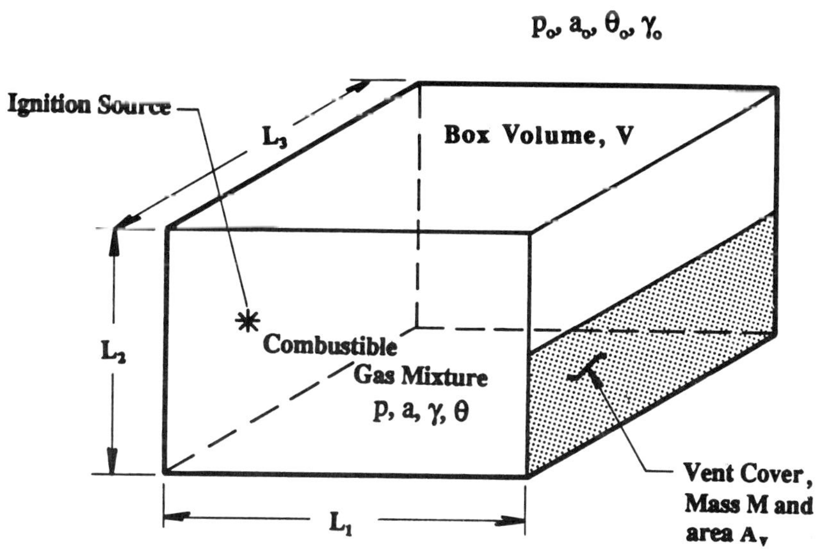

Fig. 3-1. Schematic for vented chamber containing combustible gas mixture.

(Note that the vent cover is represented by its mass and presented area, and not an opening or failure pressure). The most compact way of presenting the physical parameters is in tabular form, together with appropriate symbols and physical dimensions. Table 3-1 is such a listing, with parameters describing the chamber listed first, then ambient conditions, followed by unburned fuel-air properties, and then by combustion (or explosion) properties.

A possible set of dimensionless groups which can be obtained from this long list of physical parameters, using methods in Baker, et al (1973), appears in Table 3-2. The general procedure used was to obtain dimensionless forms of as many as possible of the dimensional quantities in Table 3-1, and to use a bar over the dimensional quantity to mean a corresponding dimensionless term.

Because we have included a large number of properties which could affect this scaling, there are of course a large number of dimensionless terms in Table 3-2. To concentrate on the more important ones, let us restrict our considerations to model-prototype situations with identical geometries, initial ambient conditions, and type and stoichiometry of gas mixture. Let us use the notation of λ with a subscript for a scale factor for a given physical quantity, meaning the ratio of the specific physical quantity in the model divided by the same physical quantity in the prototype (full-scale). A λ without a subscript means a length scale factor. So, our restrictions can then be written as:

From $$\lambda_{Av} = \lambda^2 \tag{5}$$

From $$\lambda_{L_1} = \lambda_{L_2} = \lambda_{L_3} = \lambda \tag{6}$$

$$\lambda_{a_o} = \lambda_{p_o} = \lambda_{\gamma_o} = \lambda_{\theta_i} = \lambda_{\theta} = \lambda_{K_d} = 1 \tag{7}$$

Various Pi terms in Table 3-2 then require that a number of parameters for both the unburned and burned gases be the same at any scale. These can be expressed as:

$$\lambda_a = \lambda_{\rho_u} = \lambda_{\rho_b} = \lambda_{S_u} = \lambda_{S_b} = \lambda_{\gamma_u} = \lambda_{\gamma_b} = \lambda_{\theta_b} = \lambda_{\theta_f} = \lambda_{\beta_u} = 1 \tag{8}$$

Let us use the symbols in Table 3-2 to then cast the scaling law in a reduced form showing scaled output quantities as functions of the remaining scaled quantities in the table. This gives:

TABLE 3-1 List of Physical Parameters for Vented Gas Deflagrations

Description	Symbol	Dimensions
Chamber		
Volume	V	L^3
Vent Area	A_v	L^2
Dimensions	L_1, L_2, L_3	L
Mass/area for vent cover	$m = M/A_v$	FT^2/L^3
Ambient Air Conditions		
Pressure	p_o	F/L^2
Sound speed	a_o	L/T
Temperature	θ_o	θ
Ratio of specific heats	γ_o	-
Unburned Fuel-Air Mixture		
Total energy	E	FL
Stoichiometric ratio	ϕ	-
Sound speed	a	L/T
Initial temperature	θ_i	θ
Energy release rate	$\dot{E}$	FL/T
Kinematic viscosity	ν_u	L^2/T
Thermal expansion coefficient	β_u	$1/\theta$
Ratio of specific heats	γ_u	-
Gravity	g	L/T^2
Density	ρ_u	FT^2/L^4
Combustion/Explosion Products		
Flame temperature	θ_f	θ
Burned gas temperature	θ_b	θ
Kinematic viscosity	ν_b	L^2/T
Explosion pressure	p	F/L^2
Gas specific impulse	i_g	FT/L^2
Reverberation time	t_r	T
Time	t	T
Vent time	t_v	T
Eddy size	L_e	L
Burned gas density	ρ_b	FT^2/L^4
Vent discharge coefficient	K_d	-
Ratio of specific heats	γ_b	-
Velocity of burned gas	S_b	L/T
Flame speed	S_f	L/T
Pressure rate	$\dot{p}$	F/L^2T

TABLE 3-2 Possible Dimensionless Set for Vented Gas Deflagrations

No.	Symbol	Term	No.	Symbol	Term
π_1	$\overline{A}_v$	$(A_v / V^{2/3})$	π_{15}	$\overline{S}_f$	(S_f / a_o)
π_2	$\overline{L}_i$	$(L_i / V^{1/3})$	π_{16}	$\overline{S}_b$	(S_b / a_o)
π_3	$\overline{m}$	$(M a_o^2 / p_o V^{1/3} A_v)$	π_{17}	$\overline{g}$	$(g V^{1/3} / a_o^2)$
π_4	$\overline{E}$	$(E / p_o V)$	π_{18}	$\overline{\rho}_u$	$(\rho_u a_o^2 / p_o)$
π_5	ϕ	ϕ	π_{19}	$\overline{\theta}_f$	(θ_f / θ_i)
π_6	$\overline{\dot{E}}$	$(\dot{E} / p_o a_o V^{2/3})$	π_{20}	$\overline{\theta}_b$	(θ_b / θ_i)
π_7	$\overline{a}$	(a / a_o)	π_{21}	$\overline{\nu}_b$	$(\nu_b / a_o V^{1/3})$
π_8	$\overline{\nu}_u$	$(\nu_u / a_o V^{1/3})$	π_{22}	$\overline{p}$	(p / p_o)
π_9	$\overline{\beta}_u$	$(\beta_u \theta_i)$	π_{23}	$\overline{i}_g$	$(i_g a_o / p_o V^{1/3})$
π_{10}	γ_o	γ_o	π_{24}	$\overline{t}_r$	$(t_r a_o / V^{1/3})$
π_{11}	γ_u	γ_u	π_{25}	$\overline{t}_v$	$(t_v a_o / V^{1/3})$
π_{12}	γ_b	γ_b	π_{26}	K_d	K_d
π_{13}	$\overline{L}_e$	$(L_e / V^{1/3})$	π_{27}	$\overline{\dot{p}}$	$(\dot{p} V^{1/3} / p_o a_o)$
π_{14}	$\overline{\rho}_b$	$(\rho_b a_o^2 / p_o)$	π_{28}	$\overline{t}$	$(t a_o / V^{1/3})$

$$\bar{p} = f_1(\bar{A}, \bar{L}, \bar{m}, \bar{E}, \bar{\dot{E}}, \bar{v}_u, \bar{v}_b, \bar{t}, \bar{L}_e, \bar{g})$$

$$\bar{\dot{p}} = f_2(\bar{A}, \ \ldots, \bar{g})$$

$$\bar{i}_g = f_3(\bar{A}, \ \ldots, \bar{g})$$

$$\bar{t}_r = f_4(\bar{A}, \ \ldots, \bar{g})$$

$$\bar{t}_v = f_5(\bar{A}, \ \ldots, \bar{g}) \qquad (9)$$

If we look a bit further, we see that $\pi_3 = \bar{m}$ is satisfied if we simply let

$$\lambda_m = \lambda, \qquad (10)$$

which is exactly how mass per unit area will scale if the vent covers are scaled geometrically and made of the same materials. Geometric scaling also satisfies $\bar{A}$ and $\bar{L}$, as we have said before. They were included in (9) only to show that this scaling must be maintained. Satisfying these two parameters, plus the invariance of burned gas properties, also gives Bartknecht's or Bradley and Mitcheson's scaling laws stated in Equations 1-4.

Term $\pi_4 = \bar{E}$ states now that

$$\lambda_E = \lambda^3 \qquad (11)$$

For the same fuel and stoichiometry, this is automatically satisfied, because the total energy is proportional to the volume of the mixture. Term $\pi_6 = \bar{\dot{E}}$ requires that

$$\lambda_{\dot{E}} = \lambda^2 \qquad (12)$$

This is consistent with general time scaling from $\pi_{28} = \bar{t}$, which requires that

$$\lambda_t = \lambda_{t_r} = \lambda_{t_v} = \lambda. \qquad (13)$$

Following this reasoning, we can reduce Equation (9) to:

$$\bar{p} = f_1(\bar{v}_u, \bar{v}_b, \bar{t}, \bar{L}_e, \bar{g})$$

$$\bar{\dot{p}} = f_2(\bar{v}_u, \bar{v}_b, \bar{t}, \bar{L}_e, \bar{g})$$

$$\bar{i}_g = f_3(\bar{v}_u, \bar{v}_b, \bar{t}, \bar{L}_e, \bar{g}) \qquad (9a)$$

The terms remaining on the right in Equation (9a) are all, with one exception, ones which cause problems in scaling. (The scaled time $\bar{t}$ is _not_ a problem, but simply tells us how time must scale in the scaling law.) Terms $\pi_8 = \bar{\nu}_u$ and $\pi_{21} = \bar{\nu}_b$ require that kinematic viscosity of both unburned and burned gas mixtures must scale as

$$\lambda_{\nu_u} = \lambda_{\nu_b} = \lambda \tag{14}$$

But, because both mixtures are the _same_ in model and full-scale situations, the kinematic viscosities will in fact be _unchanged_, or

$$\lambda_{\nu_u} = \lambda_{\nu_b} = 1\,. \tag{15}$$

So, we cannot satisfy these two Pi terms.

The scaled eddy size $\pi_{13} = \bar{L}_e$ is required by geometric similarity to scale by the length scale factor, or

$$\lambda_{L_e} = \lambda\,. \tag{16}$$

In fact, the physics of flame-folding and eddying may be nearly independent of the scale of the experiment, so

$$\lambda_{L_e} = 1\,. \tag{17}$$

Finally, term $\pi_{17} = \bar{g}$ fixes the relation

$$\lambda_g = 1/\lambda\,. \tag{18}$$

But, on earth, independent of the scale of testing,

$$\lambda_g = 1\,. \tag{19}$$

and g _must_ go out of scale for tests at different geometric scales. This distortion indicates that buoyant effects are not properly scaled, because they are strongly affected by gravity.

The scaling law does _not_, of course, tell us how _strongly_ distortions in these terms which cannot be properly scaled affect the desired response functions in Equation (9a). The law only states that, for complete similitude, _all_ terms in those equations _should be_ invariant--and that is only possible in a strict sense by testing _only_ to the desired scale. (In another way, we are saying that we can only reconcile Equations (14) and (15), or (16) and (17), or (18) and

(19), by requiring the length scale factor to be <u>one</u>.) This result is a very common one in similitude analysis of complex problems. It simply tells us that we should, if possible, run carefully-controlled tests or analyses where <u>only</u> the dimensionless group in question is varied, and response parameters are measured.

2. <u>Scaling of blast waves from reactive gas explosions</u>

If a deflagrating or detonating gas phase chemical reaction is confined, there is by definition no external blast wave, so the question of scaling of such waves is moot. But, if the reaction ruptures the containing structure or vessel catastrophically, or if the reaction is a UVCE, then external blast waves can develop.

Scaling laws for these classes of explosions have already been developed by Baker, et al (1983) and Oyez (1984). So, we merely state a general law of this type, without proof. Many fewer parameters are needed to describe the blast source than in the law for scaling of vented gas deflagrations, because the physics of blast wave development and transmission through the air is simpler than the processes in vented gas deflagrations.

This scaling law, as stated in Baker, et al (1983) and Oyez (1984) is:

$$\left.\begin{aligned} \bar{P} &= \left(\frac{p}{p_o}\right) \\ \bar{t}_a &= \left(\frac{t_a a_o p_o^{1/3}}{E^{1/3}}\right) \\ \bar{t}_d &= \left(\frac{t_d a_o p_o^{1/3}}{E^{1/3}}\right) \\ \bar{i}_s &= \left(\frac{i_s a_o}{E^{1/3} p_o^{2/3}}\right) \end{aligned}\right\} \begin{aligned} &= f_k\left[\left(\frac{p_1}{p_o}\right), \frac{R p_o^{1/3}}{E^{1/3}}, \gamma_1, \left(\frac{a_1}{a_o}\right), \left(\frac{\dot{E}}{E^{2/3} a_o p_o^{1/3}}\right), \ell_i\right] \\ &= f_k\left(\bar{p}_1, \bar{R}, \gamma_1, \bar{a}_1, \bar{E}, \ell_i\right) \end{aligned} \tag{20}$$

Each dimensionless parameter on the left of Equation (20) is a scaled blast wave property. This way of presenting a scaling law indicates that each such scaled property is some different (and unknown) function f_k of the six dimensionless ratios within the brackets on the right. All input physical quantities are described in Table 3-3, with dimensions in a Force-Length-Time (FLT) system.

TABLE 3-3 Physical Parameters and Their Dimensions for Gas Blast Scaling Law

Symbol	Dimensions	Description
p	F/L^2	Blast wave overpressure
t_a	T	Time of blast wave arrival
t_d	T	Duration of blast wave
i_s	FT/L^2	Positive phase specific impulse
p_1	F/L^2	Vessel failure pressure
p_o	F/L^2	Air ambient pressure
a_1	L/T	Sound speed within vessel
a_o	L/T	Sound speed in ambient air
γ_1	-	Ratio of specific heats in vessel
R	L	Distance from center of blast source
ℓ_i	-	Length ratios (enough to completely fix geometry)
E	FL	Total source energy
$\dot{E}$	FL/T	Source energy release rate

Some of the dimensionless parameters in Equation (20) are identical to those developed many years ago by Robert Sachs [Sachs (1944)], for scaling the effects of varying ambient air properties on blast waves from high explosive sources. In particular, $\bar{p}$ and $\bar{i}_s$ are often called Sachs-scaled blast overpressure and specific impulse, and $\bar{R}$

$$\overline{R} = R\,p_o^{1/3} / E^{1/3} \tag{21}$$

is usually called Sachs-scaled distance. Other terms sometimes used for $\bar{R}$ are energy-scaled distance or reduced distance.

Later in this chapter, this scaling law will be used to present results of analyses and/or test data for blast waves from gas explosions, because it allows the presentation of many results in very compact form.

C. <u>Basic properties of slow burns in gaseous systems</u>

Thermochemistry and gasdynamics of the slow mode of combustion have been described in many books such as Glassman (1977), Strehlow (1984), etc. Here, emphasis is on basic properties of the combustion process which are of interest to safety engineering.

1. <u>Heat of combustion</u>

Heat of combustion is probably the most important single parameter which defines the explosion potential of any chemical and is dependent only on the composition of the reactants. It is the heat released during the exothermic chemical reaction and serves as the energy source to perform destructive work in an accidental explosion. Usually, heat of combustion is referred to that at constant pressure and denoted by "H_c". The values of the heat of combustion for common hydrocarbon fuels are listed in Table 3-4.

Although standard computer codes are readily available for the calculation of H_c, it is still believed to be helpful to present a simple procedure to allow quick estimation of explosion potential, since it is a straightforward process with applicable accuracy.

According to the first law of thermodynamics, the heat changes which proceed from elements to combustion products by either Path A or Path B (i.e., B_1 + B_2) must be the same, as shown in the sketch.

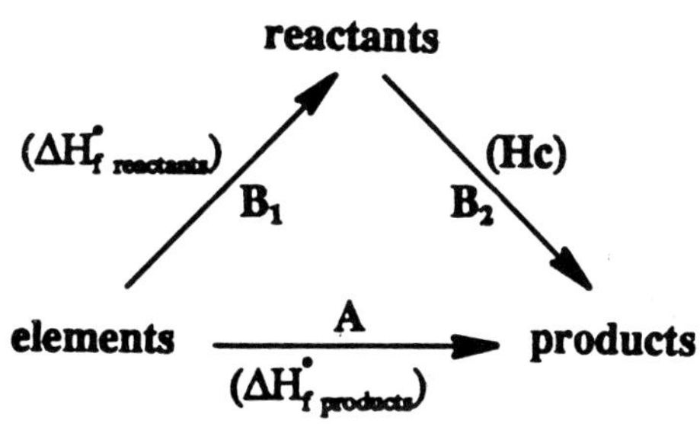

Thus, the heat of combustion can be calculated by simple difference of heat changes, i.e.,

$$H_c = \Delta H^\circ_{f\ products} - \Delta H^\circ_{f\ reactants} \tag{22}$$

where ΔH°_f is the heat of formation (or enthalpy of formation) at standard conditions (298 K and 1 atmosphere) which can be found from thermochemical tables. Heat of formation is the heat of reaction through which a given substance is formed from chemical elements. Note that heat of formation of an element at standard conditions equals zero. Heats of formation for common hydrocarbon fuels and their products are listed in Table 3-5.

Remarks:

i) Heat of combustion of a fuel is, in general, referred to the complete or stoichiometric oxidation, i.e., the heat of reaction when a general CHONS fuel reacts with O_2 and forms CO_2, H_2O, N_2 and SO_2 as products.

TABLE 3-4. Heat of Combustion of Selected Fuels*

Substances	Formula	Molecular Weight	$-\Delta H_c$ (1 atm, 298 K)(LHV)	
			MJ/mol	MJ/kg
Methane	CH_4 (g)	16	0.8826	55.164
Ethane	C_2H_6 (g)	30	1.5425	51.416
Propane	C_3H_8 (g)	44	2.2036	50.082
n-Butane	C_4H_{10} (g)	58	2.8806	49.665
Isobutane	C_4H_{10} (g)	58	2.8614	49.334
n-Pentane	C_5H_{12} (g)	72	3.5100	48.750
	C_5H_{12} (l)	72	3.4895	48.465
n-Hexane	C_6H_{14} (l)	86	4.1443	48.189
n-Heptane	C_7H_{16} (l)	100	4.8146	48.146
n-Octane	C_8H_{18} (l)	114	5.4544	47.846
Decane	$C_{10}H_{22}$ (l)	142	6.7419	47.478
Ethylene	C_2H_4 (g)	28	1.3884	49.586
Propylene	C_3H_6 (g)	42	2.0525	48.868
n-Butylene	C_4H_8 (g)	56	2.7206	48.581
Isobutylene	C_4H_8 (g)	56	2.7098	48.390
Pentylene	C_5H_{10} (g)	70	3.3783	48.261
	C_5H_{10} (l)	70	3.3638	48.055
Hexene	C_6H_{12} (l)	84	3.9885	47.483
Acetylene	C_2H_2 (g)	26	1.3063	50.244
Propyne	C_3H_4 (g)	40	1.9474	48.684
Butine	C_4H_6 (g)	54	2.5995	48.140
Pentyne	C_5H_8 (g)	68	3.2576	47.906
Cyclopentane	C_5H_{10} (l)	70	3.2809	46.870
Cyclohexane	C_6H_{12} (l)	84	3.9266	46.745
Ethylene oxide	C_2H_4O (l)	44	1.2649	28.748
Propylene oxide	C_3H_6O (l)	58	1.888	32.550
Alcohol	C_2H_6O (l)	46	1.3717	29.819
Ether	$C_4H_{10}O$ (l)	74	2.7287	36.874
Acetone	C_3H_6O (l)	58	1.7870	30.811
Benzene	C_6H_6 (l)	78	3.2755	41.993
Propadiene	C_3H_4 (g)	40	1.9612	49.03
Butadiene	C_4H_6 (g)	54	2.5375	46.99**
Hydrogen	H_2 (g)	2	0.244	122.051

TABLE 3-4. (Continued)

Substances	Formula	Molecular Weight	$-\Delta H_c$ (1 atm, 298 K)(LHV)	
			MJ/mol	MJ/kg
Ammonia	NH_3 (g)	17	0.3164	18.61**
Aluminum	Al (s)	27	0.8389	31.07
Sulfur	S (s)	32	0.2963	9.26
MAPP***	$C_3H_{5.08}$ (g)	41	1.820	44.380
TNT****	$C_7H_5N_3O_6$ (s)	227	3.4363	15.138

* From "Handbook of Chemical Engineering" by Publishing House of Chemical Industry, China, 1989.

** From Eichler and Napadensky (1977).

*** Mixture of Methyl-acetylene 51.0%, propane 26.0%, and propadiene 23.0%.

**** Heat of explosion for TNT = 4.187 MJ/kg.

TABLE 3-5. Heat of Formation for Selected Fuels and Combustion Products*

Substances	Formula	Molecular Weight	ΔH_f(1 atm, 298 K) KJ/mol
Oxygen Atom	O (g)	16	249.35
Hydrogen Atom	H (g)	1	218.12
Hydroxide Radical	OH (g)	17	42.03**
Water Vapor	H_2O (g)	18	-242.00
Water	H_2O (l)	18	-286.06
Nitrogen Atom	N (g)	14	473.04
Nitrogen Oxide	NO (g)	30	90.31
Nitrogen Dioxide	NO_2 (g)	46	33.08
Ammonia	NH_3 (g)	17	-46.14
Nitric Acid	HNO_3 (g)	63	-135.16
	HNO_3 (l)	63	-173.34
Ammonium Nitrate	NH_4NO_3 (s)	80	-365.82
Ammonium Chloride	NH_4Cl (s)	53	-314.65
Ammonium Perchlorate	NH_4ClO_4 (s)	117	-290.04**
Sulphur Dioxide	SO_2 (g)	64	-297.03
	SO_2 (l)	64	-320.72
Sulfuretted Hydrogen	H_2S (g)	34	-20.64
Carbon Atom	C (g)	12	718.90
Graphite	C (s)	12	0
Diamond	C (s)	12	1.90
Carbon Monoxide	CO (g)	28	-110.60
Carbon Dioxide	CO_2 (g)	44	-393.79
Chlorine Atom	Cl (g)	35	121.77
Chlorine Hydride	HCl (g)	36	-92.37
Bromine Atom	Br (g)	80	111.96
Bromine Hydride	HBr (g)	81	-36.18**
Fluorine Atom	F (g)	19	79.05
Fluorine Hydride	HF (g)	20	-271.32
	HF (l)	20	-300.00
Methane	CH_4 (g)	16	-74.90
Ethane	C_2H_6 (g)	30	-84.74
Propane	C_3H_8 (g)	44	-103.92
n-Butane	C_4H_{10} (g)	58	-124.81
Isobutane	C_4H_{10} (g)	58	-131.69
n-Pentane	C_5H_{12} (g)	72	-146.54
	C_5H_{12} (l)	72	-173.17

TABLE 3-5. (Continued)

Substances	Formula	Molecular Weight	ΔH_f (1 atm, 298 K) KJ/mol
Hexane	C_6H_{14} (g)	86	-167.31
	C_6H_{14} (l)	86	-198.97
Heptane	C_7H_{16} (g)	100	-187.95
	C_7H_{16} (l)	100	-224.55
Octane	C_8H_{18} (g)	114	-208.60
	C_8H_{18} (l)	114	-250.13
Nonane	C_9H_{20} (g)	128	-228.70**
	C_9H_{20} (l)	128	-275.08**
Decane	$C_{10}H_{22}$ (g)	142	-249.30**
	$C_{10}H_{22}$ (l)	142	-300.61**
Ethylene	C_2H_4 (g)	28	52.32
Propylene	C_3H_6 (g)	42	20.43
Butylene	C_4H_8 (g)	56	1.17
Isobutylene	C_4H_8 (g)	56	-13.98
Pentylene	C_5H_{10} (g)	70	-20.94
Hexene	C_6H_{12} (l)	84	-41.57**
Heptylene	C_7H_{14} (l)	98	-63.88**
Octylene	C_8H_{16} (l)	112	-86.23**
Nonene	C_9H_{18} (l)	126	-108.50**
Acetylene	C_2H_2 (g)	26	226.89
Propyne	C_3H_4 (g)	40	185.57
1-Butine	C_4H_6 (g)	54	166.22
2-Butine	C_4H_6 (g)	54	148.09
1-Pentyne	C_6H_8 (g)	68	144.45
2-Pentyne	C_5H_8 (g)	68	128.96
1,2 Butadiene	C_4H_6 (g)	54	165.60
1,3 Butadiene	C_4H_6 (g)	54	112.00
Cyclohexane	C_6H_{12} (g)	84	-123.22
	C_6H_{12} (l)		-156.34
Benzene	C_6H_6 (g)	78	82.90
	C_6H_6 (l)		48.99
Alcohol	C_2H_6O (g)	46	-218.69
	C_2H_6O (l)		-277.81
Ether	$C_4H_{10}O$ (l)	74	-272.99

TABLE 3-5. (Continued)

Substances	Formula	Molecular Weight	ΔH_f(1 atm, 298 K) KJ/mol
Ethylene Oxide	C_2H_4O (g)	44	-71.18
	C_2H_4O (l)		-97.56
Propylene Oxide	C_3H_6O (l)	58	-120.59
MAPP	$C_3H_{5.08}$ (g)	41	113.98**
Aluminum Oxide	Al_2O_3 (s)	102	-1676.89
Potassium Chlorate	$KClO_3$ (s)	122	-391.48
Potassium Perchlorate	$KClO_4$ (s)	138	-433.77

* From "Handbook of Chemical Engineering" by Publishing House of Chemical Industry, China, 1989.

** From: NBS Circular 500, "Selected Values of Chemical Thermodynamic Properties."

For instance, complete, or stoichiometric oxidation of paraffin hydrocarbons can be illustrated by the following reaction equation:

$$C_nH_{2n+2} + (1.5n+0.5)\,O_2 \rightarrow nCO_2 + (n+1)\,H_2O \qquad (23)$$

The volume percent of stoichiometric mixture (C_{st}) of hydrocarbons in air is

$$C_{st} = \frac{100}{1+4.773\,(1.5n+0.5)} \qquad (24)$$

where n = positive integer. Practically, complete oxidation does not occur because of an insufficient supply of oxygen and the dissociation of combustion products. In these cases, heat of combustion is determined according to the practical or equilibrium composition of reaction products. It is, of course, significantly less than the heat release at complete oxidation.

ii) For gaseous fuels, the composition is ordinarily given by the volume percent of the component gases and thus, H_c has the unit of kJ/mole of fuel. For solid and liquid fuels, the ultimate is given on a weight or mass percent of the elements and the heat of combustion is by per unit weight or mass of fuel, i.e., kJ/kg of fuel. Nevertheless, heat of combustion on a molar basis can be easily converted to a mass basis or vice versa. Dividing H_c in kJ/mol by the molecular weight of the fuel, then multiplying by 1000, will convert H_c in kJ/mol to H_c in kJ/kg.

iii) Since water can be assumed to be either a liquid or a gas at the end state, there are two values of H_c that are normally reported. One is the high value (HHV) when liquid water is formed. The other is the low value (LHV) when gaseous water is the end product. The difference between the high and low values of the heat of combustion for any substance is just the heat of vaporization of water which equals 44.0 J/mol at 25°C. The low value should be used in the calculation of H_c because these combustion reactions produce steam rather than water, and condensation occurs relatively slowly.

iv) The heat of explosion of a propellant or an explosive is determined experimentally by exploding or burning the material in a bomb filled with inert gas, e.g., N_2, atmosphere. The final state attained in the experiment is not the fully oxidized state of the substance, because these substances are usually oxygen deficient. As the result, heat of explosion is in magnitude less than the heat of combustion of the same substance which is measured in a bomb filled with pure oxygen; e.g., for TNT, heat of explosion H_e is about 4,500 kJ/kg, while the heat of combustion H_c is about 12,500 kJ/kg.

v) Combustion reactions are highly exothermic and thus H_c is a negative number according to the definition that heat rejection is negative. However, it should be noticed that it is reported as a positive number in some of the other sources.

The adiabatic flame temperature is the highest temperature that can be achieved by the combustion products. By definition, if there is no work done, there are no heat transfer losses. In other words, if all the heat of reaction remains in the products, the temperature of the combustion products achieved is called adiabatic flame temperature, which is usually denoted by T_f. For most stoichiometric fuel-air mixtures, T_f is of the order of 2,000-2,500 K, while for fuel-oxygen mixtures it is about 3,000-3,500 K. At the above temperature range, dissociation of molecules can no longer be neglected. The calculation of T_f, even without taking the dissociation reactions into account, is not a straightforward matter but an iteration procedure.

An iteration procedure has been developed to calculate H_c and T_f based on the equilibrium compositions of the combustion products, providing that equilibrium constants for most gases in the products are given in the JANAF Tables. Standard computer codes are available for such equilibrium thermodynamic calculations with the composition of reactants as the input since both H_c and T_f are basic properties of a combustion reaction which depend only on the composition of the reactants. One commonly used computer program appears in NASA SP-273 by Gordon and McBride (1971, revised 1976). It can be purchased from the National Technical Information Service in Washington D.C.

2. Laminar burning velocity

Laminar burning velocity is a fundamental quantity of a flame which depends only on initial conditions. It characterizes the reactivity or reaction rate of the premixed gaseous mixture. The larger the laminar burning velocity, the higher the reactivity of the reactive mixture. By definition, it is the propagation velocity of the flame front with respect to unburned gas or the volume of unburned gas consumed by the flame per unit time. One must distinguish laminar burning velocity $\underline{b}$, from flame speed $\underline{v}$, which is the propagation velocity of the flame front with respect to the laboratory coordinate. Flame speed is usually orders of magnitude larger than laminar burning velocity because of the flow velocity of the unburned gas ahead of the flame.

Physical processes responsible for flame propagation are heat conduction and the diffusion of active species (i.e., free radicals) from the reaction zone into the unreacted region in front of the reaction zone. By the simultaneous solution of both the heat conduction equation and the mass diffusion equation along with gasdynamic equations numerically, standard computer codes are available [Smoot, et al (1976) and Tsatsaronis, G. (1978)] for the determination of the laminar burning velocity in a given mixture. Nevertheless, a major difficulty lies in the lack of accurate chemical rate laws and the transport coefficients of the various molecules involved in most of the common reactions. On the other hand, experimental measurements of laminar burning velocity are rather straightforward, thus, the parameter b can be considered as an experimentally determined constant for a given reactive gas. Unfortunately, an accuracy of ±20 percent is about the best that can be expected at the present time [Andrews and Bradley (1972)]. A comprehensive review of the experimental measurement of b is given by Linnett (1953).

A more complete collection of data on laminar burning velocities for some of the commonly used hydrocarbons are listed in Appendix C according to NFPA68.

It can be seen from the table in Appendix C that laminar burning velocity is of the order of 0.5 m/s for most stoichiometric hydrocarbon-air mixtures while that for stoichiometric fuel-oxygen mixtures is an order of magnitude higher. This is because fuel-O_2 mixtures are more energetic mixtures with much higher flame temperatures resulting in much faster chemical reactions. Note that hydrogen and acetylene have unusually high burning velocities due to high flame temperatures and high diffusivity associated with the light hydrogen molecules.

Laminar burning velocity and other properties of a flame are usually presented in terms of the equivalence ratio of the fuel, ϕ , which is defined using equation (25),

$$\phi = \frac{f/a}{(f/a)_{stoich}} \tag{25}$$

where f is the fuel concentration and a is the concentration of oxidizer. The subscript "stoich" represents the f/a for a stoichiometric fuel-oxidizer mixture. By normalizing the fuel-oxidizer concentration ratio with the stoichiometric ratio, the value of ϕ gives a straightforward presentation of the oxygen supply in the system. When $\phi < 1$, the mixture is said to be fuel lean which means there is excess oxygen in the system. When $\phi > 1$, the mixture is said to be fuel rich which means there is excess fuel in the system. At $\phi = 1$, the combustion reaction approaches completion and the system is called a stoichiometric mixture.

Laminar burning velocity depends on stoichiometry of the mixture. The relation of b versus ϕ is an inverse "U" shaped curve with b decreasing for off-stoichiometric mixtures to very low values at the lean and rich limits of flammability. Most of the flames in gaseous mixtures show a maximum flame temperature and burning velocity at a slightly rich mixture rather than ϕ = 1 when chemical equilibrium is taken into account.

3. Flammability limits

Unlike T_f and b, flammability limit is not a unique physical property of the system but rather is affected by many factors other than the initial conditions. Nevertheless, flammability limit is of considerable importance to industrial safety, because it indicates the ability of the mixture to support flame propagation.

Flammability limits are defined as the limiting fuel concentrations between which a flame will propagate whereas outside of this range a sustained self-propagation of a flame is not possible, i.e., the flame will not propagate after the ignition source is removed.

The United States Bureau of Mines has established an experimental method to determine flammability limits of gases and vapors [Coward and Jones (1952)]. In this technique, a 51 mm internal diameter glass tube 1.8 m long is mounted vertically and closed at the upper end with the bottom end opened to the atmosphere. The gas mixture to be tested is placed in the tube and ignited at the lower end by an electric spark. The limits are determined by visual observation as to whether a flame is seen to propagate up the glass tube or not.

The U.S. Bureau of Mines has been accumulating flammability data over the years from various laboratories throughout the world, and these data are published by Coward and Jones (1952) and updated by Zabetakis (1965). These publications have been the main sources of flammability data so far.

Some flammability data for commonly used combustible gases are listed in Tables 3-6 and 3-7.

Note that the flammability limit data are not much different from the explosion limit data and the latter is not well defined. Also, the lean limit of the same fuel mixed with air or oxygen rarely differs, because the excess oxygen in the lean condition has the same thermophysical properties as nitrogen.

Flammability limits of silane, one of the important material gases in the semiconductor industries, were determined in a cylindrical vessel with diameter of 100 mm and 100 mm long [Horiguchi et al (1987)]. The lower flammability limit is 1.37% in silane-air and 0.8% in silane-hydrogen-air mixtures at room temperature and atmospheric pressure. The upper limit could not be measured because ignition occurred spontaneously during the mixing process.

An empirical relation has been concluded from a large amount of data available. In general, the lean flammability limit (LFL), which is most important in industrial safety, for hydrocarbon gases appears at about 55 percent of the stoichiometric fuel concentration (in volume percent), while the rich limit (UFL) occurs at about 330 percent of stoichiometric [Mullins and Penner

TABLE 3-6. Explosion Related Parameters for Gases [TNO, Anon., (1980b)]

Gas	Explosion Limits (vol. %)	Flash Point (°C)	Autoignition Temperature (°C)	Shock Wave Reactivity
Acetaldehyde	4-57	-50	140	average
Acetylene	1.5-100	-84	300	high[1]
Ammonia	15-28	-33	630	low
1,3-Butadiene	1.1-12.5	-85	415	average
n-Butane	1.5-8.5	-138	360	average
1-Butene	1.6-10	-185	380	average
Carbon Monoxide	12.5-74.2	-191	605	low
Dimethylamine	2.8-14.4	-18	400	average
Ethyl Chloride	3.8-15.4	-58	510	low
Ethane	3.0-13.5	-89	510	average
Ethene	2.7-34	-104	425	average
Ethylene Oxide	3-100	-18	440	high
Formaldehyde	7-73	-19	430	high
Hydrogen Sulphide	4.3-46	-60	270	high
Methane	5-15	-161	540	low
Methyl Bromide	8.6-20	4	530	low
Methyl Chloride	10.7-17.2	-24	625	low
Propane	2.1-9.5	-187	465	average
Propene	2-11.7	-48	455	average
Vinyl Chloride	4-29	-77	470	average

[1] In view of the high reactivity of acetylene, it is advisable for this gas to use only the top line of the "high reactivity" range.

TABLE 3-7. Flame Properties for Some Selected Fuel-Air Mixtures [Baker, et al. (1983)]

Fuel	Maximum Value		Minimum Value			LEL[4] % Fuel	UEL[4] % Fuel
	θ_f[1] °K	S_u[2] m/sec	E min[2] millijoules	$d_{\|\|}$[3] mm	AIT[4] °K		
Hydrogen	2400[5]	2.70	0.018[5]	0.55	673	4.0	75.0
Carbon Monoxide (wet)	2370	0.33[5]	-	-	826[6]	12.5	74.0
Methane	2230	0.34	0.280	2.50	713	5.0	15.0
Acetylene	2610	1.40	-	0.55	578	2.5	100.0
Ethylene	2395	0.63	-	1.25	763	2.7	36.0
Ethane	2170[5]	0.44	0.250	2.00[5]	788	3.0	12.4
Propane	2285	0.39	0.260	2.10	723	2.1	9.5
n-Butane	2170[5]	0.35	0.260	2.20	678	1.8	8.4

[1] Calculated using JANAF, Strechlow (1981)

[2] NACA (1959)

[3] Potter (1960)

[4] Zabetakis (1965)

[5] Lewis and Von Elbe (1961)

[6] Alroth, et al. (1976)

(1959)]. One empirical observation [Bodurtha (1980)] is that for many hydrocarbons the lean limit of fuel (volume %) multiplied by its heat of combustion in kJ/mol, is approximately 4.35 x 10^3. This means that the flame temperature is approximately constant at extinction for all hydrocarbon fuels, which is so low that the kinetic mechanism changes.

For determining the lean limit of mixed fuel, the following equation was proposed by Le Chatelier in 1898, which has worked quite well for most hydrocarbons:

$$L_m = \frac{100}{\frac{C_1}{L_1} + \frac{C_2}{L_2} + \ldots + \frac{C_n}{L_n}} \tag{26}$$

where C_1, C_2, ... C_n are the concentrations (in volume %) of each constituent fuel in the mixture, respectively; L_1, L_2, ... L_n are the lean flammability limits (in volume %) of each constituent fuel in the mixture, respectively; L_m is the lean flammability limit (in volume %) of the mixture.

There exists no good theory for predicting flammability limits, nor a standard experimental technique to measure them as yet. Nevertheless, it is useful to discuss briefly the various factors which influence the limits.

The effect of buoyancy is perhaps the most important one. It is known that upward propagation has wider flammability limits than downward propagation and that of horizontal propagation is in between. This is due to the very low burning velocity in the near limit mixture which could be exceeded by the buoyancy-induced flow velocity and hence the flame cannot burn downwards.

Regarding the dependence of the limits on the initial thermodynamic states of the mixture, the increase of initial temperature of the mixture will widen the flammability limits in general, since high initial temperature will facilitate the flame propagation by reducing the energy required to heat the gas in preheating zone. As for pressure dependency, there are no general rules available. For hydrocarbons, the lean limit appears to be very insensitive to initial pressures above atmospheric. For instance, the lean limit for methane is 5 percent at atmospheric pressure and stays constant at 5 percent up to the pressure of 100 atmospheres. On the other hand, the rich limits of hydrocarbons widen dramatically at elevated initial pressures, e.g., the rich limit of methane in air increases from 15 percent at atmospheric pressure to about 40 percent at the pressure of 100 atmospheres.

The apparatus-dependent aspect of flammability limits is mostly due to various energy loss mechanisms. It has been found that as the tube becomes smaller, the combustible range becomes narrower until quenching diameter is reached. The effect of tube wall on heat loss is more severe for small tubes.

The choice of 51 mm inner diameter as the standard apparatus by U.S. Bureau of Mines is because this is the diameter at which a further increase in tube diameter causes only a slightly widening of the limits that are measured.

Finally, in the measurement of flammability limits we have assumed that the ignition source is strong enough that the influence of ignition source can be ignored. In fact, the measured limits will be narrower if the ignition source is not sufficiently strong, because near limit mixtures require very strong ignition sources to initiate the combustion reaction.

4. Autoignition Temperature (AIT)

Thermal ignition of a reactive gas is simply to heat a certain volume of the gas to a temperature in which rate of heat generation is greater than the rate of heat loss and hence, self-acceleration of reaction can occur. The relation between the rate of heat generation and heat loss is sketched below.

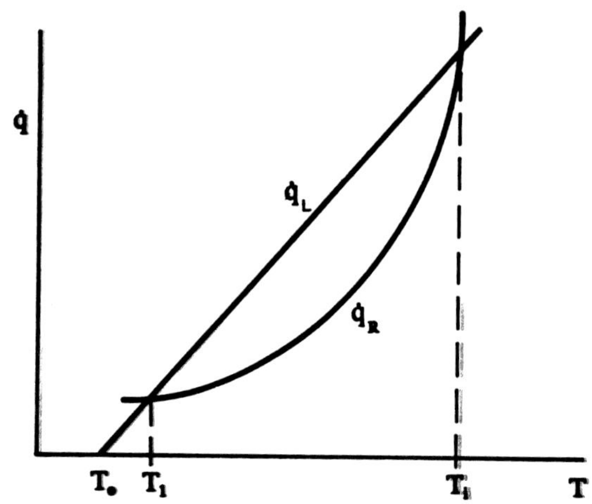

Fig. 3-2. The sketch of heat generation and heat loss.

where $\dot{q}_R$ is the rate of heat generation by chemical reactions, which is dependent on the overall reaction rate and therefore $\dot{q}_R$ vs T is an exponential curve for any reaction according to the Arrenhius law of mass action. $\dot{q}_L$ is the rate of heat loss by conduction which is proportional to the temperature difference between the reactive mixture and the surroundings and the surface to volume ratio of the gas volume. It is thus a straight line in $\dot{q}_L$ vs T plot.

For a given combustible mixture, the heat generation and heat loss rates will compete with each other as shown in the sketch. Presumably, the volume of combustible mixture is stable at initial temperature T_o. Self-heating will bring the temperature up to $T = T_1$ in which steady-state occurs, i.e., $\dot{q}_R = \dot{q}_L$. External addition of energy may bring the temperature up, but the temperature will decrease until it is back to $T = T_1$, since $\dot{q}_L > \dot{q}_R$ in the region of T_1 to T_i.

This occurs unless the addition of external energy is at a sufficiently rapid rate (must exceed $\dot{q}_L$) and then bring the temperature of the system beyond T_1, in which rate of heat production $\dot{q}_R$ again dominates over the heat loss term $\dot{q}_L$.

The critical temperature T_1, at which $\dot{q}_R$ starts to dominate and hence self-acceleration of chemical reactions occur, is defined as <u>ignition temperature</u>. We can see that the ignition temperature is not unique but dependent on the size of the heated volume, since $\dot{q}_L$, the rate of heat loss per unit volume, is proportional to the area to volume ratio.

On the other hand, there is another important concept about thermal ignition, namely <u>ignition time delay</u> or <u>induction time</u>. For an adiabatic system at any initial conditions the temperature, and hence reaction rate, change very slowly during the period of so-called induction time, then the reaction occurs almost "instantaneously" at the termination of the induction period. In other words, temperature and pressure of the system rise almost instantaneously after an induction time. This behavior is due to the high activation energy of hydrocarbon fuels. Although the induction time is derived from an adiabatic volumetric reaction, it is really a global scale that characterizes a given combustible mixture, since the reaction time is relatively short and heat loss can be ignored even when the mixture is not bounded by insulated walls.

The induction time is a strong function of temperature and increases rapidly with the decrease in temperature. Eventually, there is a temperature below which the induction time is so long that all the combustible mixture will be consumed before the explosion reaction can occur. In a given size volume (using a standard apparatus) the minimum temperature at which the mixture can attain self-acceleration conditions is called the minimum <u>autoignition temperature (AIT)</u>.

A typical example of a measurement of the ignition delay to determine the AIT is given in Fig. 3-3 for normal propylnitrate (Zabetakis, 1965). In this case, the minimum AIT was determined to be 170°C.

In regard to industrial safety, the minimum autoignition temperature gives some indication of the maximum surface temperature that should be allowed when the particular combustible mixture is present in an industrial environment.

The AIT data for some hydrocarbon fuel and air mixtures are listed in Table 3-7 with flammability and burning velocity data.

5. <u>Ignition energy and quenching distance</u>

Ignition energy and quenching distance are two other fundamental properties of a flame which are of importance to the flame propagation and safety. As illustrated by the previous sketch, the addition of an external energy is required for the ignition of any combustible mixture which is stable at ordinary temperature, i.e., to bring the mixture to ignition temperature T_1, at which $\dot{q}_R$ starts to surpass $\dot{q}_L$.

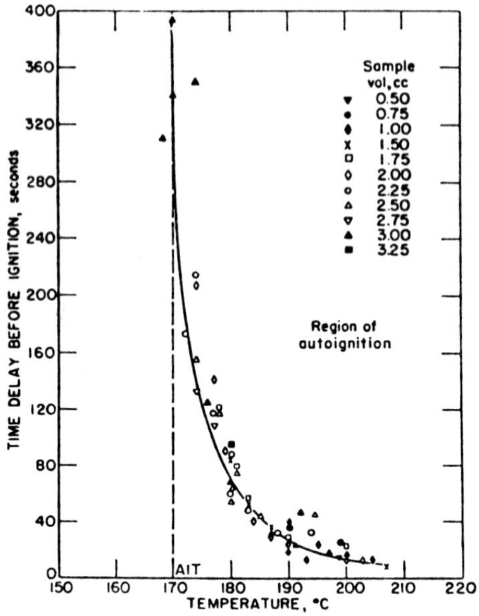

Fig. 3-3. Time delay before ignition of NPN in air at 6900 kPa in the temperature range from 150°C to 210°C [Zabetakis (1965)].

To bring about ignition of the entire volume of combustible mixture, an amount of external energy E_e, which can increase the temperature of the entire volume to T_i, is required. Alternatively, the case of more general interest is to bring about ignition in a small local volume of the mixture only and the external energy E_i required will be much smaller in this case. Thus, the ignition energy depends upon the size of the local volume of the mixture from which the combustion reaction can spread. This is referred to as the minimum flame kernel. The volume of the kernel can be estimated approximately as a slab with the thickness of the order of flame thickness (inversely proportional to laminar burning velocity), and the area of the slab $A = d_c^2$, where d_c is the flame quenching distance, which is the minimum tube diameter in which a flame cannot propagate due to excessive heat loss to the surroundings. The quenching diameter may be related to the flame thickness δ by the empirical relation $d_c \approx a\delta$, where a is a constant of the order of 40 and is not very sensitive to the properties of the mixture [Lee (1981)]. Thus the minimum ignition energy may be expressed by an approximate relation as:

$$E_i \approx a^2 \delta^3 \rho c_p (T_f - T_0) \tag{27}$$

or

$$E_i \approx \alpha \frac{T_f - T_0}{b^3} \tag{28}$$

Note that the approximate expression given above can provide only a qualitative description of E_i. Nevertheless, it can be used to explain the experimentally well demonstrated feature of the sharp U-shaped curve of E_i versus fuel concentration, since b and T_f with composition curves both have the form of an inverted U with the maximum at around the stoichiometric composition.

It should be noted that the minimum ignition energy of a given mixture is not unique but affected by many factors. First of all, types of ignition sources and the rate of energy deposit must be considered. It is still a subject of research and has been discussed in detail in many papers. Here, to simplify the matter, we assume that the external energy is deposited instantaneously. Thus, we may eliminate other parameters of the ignition source such as the rate of energy deposit, and speak of the energy addition only.

Hertzberg (1980) has presented a simple analytical model to estimate the fraction of energy of a capacitance spark that is available to initiate the combustion in the test mixture. The maximum possible is about 70% of the stored electrical energy (1/2) cv^2, where c is the capacitance and v is the input voltage, since a fraction of the deposited energy must appear as the work of the compression of the surroundings. In fact, the largest measured efficiency is only about half of the possible maximum, i.e., about 35% of the stored electrical energy. The rest of the ignition energy is probably dissipated as the heat in the external circuitry, at the electrodes or as the shock energy.

Values of the minimum energy for some fuels mixed with air are also contained in Table 3-7. Note that the minimum ignition energies for gaseous mixture are extremely small. As a matter of act, sparks with sufficient energy to ignite most hydrocarbons can be generated by ordinary static electricity, or in other words, by commonly used electric equipment. Therefore, in ordinary industrial situations, spark ignition of combustible gas or vapor and air mixtures is an ever present possibility and special precautions must be taken to reduce the possibility of ignition when gases or vapors are present in the combustible range.

Flame quenching refers to the fact that a flame cannot propagate in small tubes or transmit through a narrow gap or a small hole due to the excessive heat loss to the walls. The subject of flame quenching is of great importance in combustion safety. The quenching diameter or gap is a very important parameter in the design of electrical equipment used in a combustible environment. Standard apparatus are devised to measure the so-called <u>safety gap</u> for most of the fuel-air mixtures and extensive data are contained in The National Fire Codes (NFC), particularly National Electrical Code [NFPA70 (1990)].

In general, the quenching distance is of the order of 1 to 2 mm for common hydrocarbon fuels mixed with air, and about 10 times smaller for the same fuel mixed with pure oxygen. Again, H_2 and C_2H_2 are exceptional because they have quenching distances much smaller than other common fuels.

Likewise the minimum ignition energy also depends on various flame properties. The approximate expression of d $\simeq$ 30 $\simeq$ 50 δ indicates the U-shaped curve for the dependence of the quenching diameter on fuel composition with the minimum near stoichiometric composition. The inverse dependence of d on laminar burning velocity b also enables the influence of initial temperature and pressure on the quenching distance d to be evaluated. It should also be mentioned that turbulence increases the quenching distance since the rate of turbulent heat transfer is greater than the corresponding laminar case.

In the above discussion only heat loss has been considered. In reality, however, the cold walls may provide a sink for the free radicals in the flame as well. For further improvement, mass diffusion losses should also be included in a proper quenching distance analysis.

D. Confined and vented gaseous deflagrations

Technically, completely confined gaseous deflagrations are not explosions, because no shock waves form within the confinement and there are no external explosion effects because the confining structure or vessel is strong enough to completely contain the pressures created within the enclosure by the deflagration. But, if the deflagration pressure exceeds the strength of the enclosure, a damaging explosion will result as the enclosure fails and the pressure is suddenly released. In industrial processes which can result in gaseous deflagrations within enclosures, it is common practice to control or mitigate the effects using venting, inerting or suppression techniques. Consequently, there is voluminous literature on this topic, as there is for prediction and control of the somewhat similar physical processes in confined dust explosions. Good general references on this topic are Bartknecht (1978a and 1981), NFPA68 (1988), NFPA69 (1986), Baker, et al (1983), and Oyez (1984).

1. Unvented gas deflagrations

Deflagrative pressures for combustible gas mixtures recorded in closed vessels of spherical or nearly spherical geometry with few internal obstructions and "soft" ignition are essentially identical in character to vessel tests for suspended dust-air mixtures, as in Fig. 2-1 [see Bartknecht (1978a, 1981) and Zabetakis (1965)]. That is, pressure rises following a cubic in time until combustion is complete, then levels off at a maximum value (p_e absolute, P_m gage), after which it slowly decays because of heat transfer through the vessel. Scaling of the maximum rate of pressure rise, $(\dot{p})_{max}$, has been found to follow Bartknecht's cubic law, Equation (1). The proportionality constant for a given gas in this law is denoted K_g, with the subscript indicating gas. Strong turbulence within an enclosure, or complex internal geometry causing flame acceleration processes usually lead to greater values for $(\dot{p})_{max}$ than reported based on closed vessel testing.

Data for the closed vessel tests generally is more repeatable, and the tests are easier to run than for combustible dusts, because uniform combustible mixtures are easily achieved, and the mixtures are easy to ignite. Table 3-8

TABLE 3-8. Flammable Gases (NFPA 68)

	Flammable Range, mol %		P, bar ga		
Flammable Gas	Lit.	Measured	Lit.	Measured	K_G bar · m/sec
Methane	5 - 15	5 - 14.5	7.2	7.05	64
Ethane	3 - 12.5	2.5 - 13	-	7.8	106
Propane**	2.1 - 9.5	2.0 - 10.0	8.6	7.9	96
Butane**	1.5 - 8.5	1.75 - 8.5	8.6	8.0	92
Pentane	1.4 - 7.8	1.5 -	8.7	7.65	104
Ethylbenzene	1.0 - 7.8	0.5 -	-	6.6	94
Acetophenone	-	0.8 -	-	6.9	109
Hydrogen	4.0 - 75.6	5.0 - 72.5	7.4	6.9	659

gives some data on maximum gas pressure P_m and K_g, from NFPA68. That reference also recommends calculation of K_g for untested flammable gases by the simple proportionality equation

$$(K_g) = (K_g)_1 \left[\frac{(S_u)_2 \cdot (P_m)_2}{(S_u)_1 \cdot (P_m)_1} \right] \qquad (29)$$

Values for P_m can both be measured or both be calculated assuming constant volume heat addition to the gas in the closed vessels. But, one should not calculate one value and measure the other. Values of S_u for many gases are given in NFPA68, and are reproduced in Appendix C.

2. Vented gas deflagrations

As is true for confined dust explosions, a popular method of explosion control is passive venting through a low-strength and low-mass vent cover. Again, the four general references cited earlier cover this topic in considerable detail, along with Bradley and Micheson (1978a, 1986). Referring to Fig. 2-21, one reduces the maximum pressure within an enclosure by providing a vent with enough area A_v and low enough release pressure (P_{stat} or P_v) to limit the maximum internal pressure (P_1, P_2, or P_{red}) to a safe value. The scaled vent area

$$\overline{A}_v = (A_v / V^{2/3}) \tag{30}$$

and scaled mass of vent cover

$$\overline{m} = (Ma_o / p_o V^{1/3} A_v) \tag{31}$$

from Table 3-2 are important parameters. Bradley and Micheson's work gave scaled prediction curves for estimating P_{stat} for venting of roughly spherical vessels, Fig. 3-4 and Table 3-9. [Note that their A is defined in a slightly different way than Equation (24). Also, they do not give enough information on covered vents to calculate $\overline{m}$.] Venting nomographs have also been developed by Bartknecht (1978a and 1981), and are reproduced in NFPA68. A typical one for hydrogen is shown in Fig. 3-5.

It will be apparent from discussions later in this chapter that gaseous deflagrations in long enclosures and/or with turbulence cannot be safely vented using information such as Fig. 3-4 or 3-5. Also, as size is increased, parameters which do scale according to the simple cubic law, as described in Equations (9a) and (11) through (16) have been found to seriously affect maximum pressures in large vented enclosures. Pappas (1984) summarizes related tests, as in Fig. 3-6, and also presents empirical prediction curves, dependent on vessel volume (Fig. 3-7), based primarily on large-scale testing in Norway. But, we are very far from identifying which ones of the distortions in scale factors shown in Equation (9a) is affecting the maximum vented pressure.

E. Detonation fundamentals

Reactive gas detonation has been a subject of great interest for over a century due to its violent nature and involvement in various types of accidental explosions. The topics of major concern are the sensitivity of a given combustible mixture to the onset of detonation and the damage potential once a detonation has been initiated. The former is referred to as the initiation of detonation and the latter, the detonation parameters.

This section will start with the determination of the detonation parameters of an established detonation which are of importance in the evaluation of the possible violence level a gaseous explosion could achieve. It is well understood that once the composition of a combustible mixture is given, detonation parameters of the system can readily be calculated with considerable accuracy on the basis of thermodynamic data alone. Therefore, more emphasis will focus on the problem of detonation initiation, which plays a controlling role in the

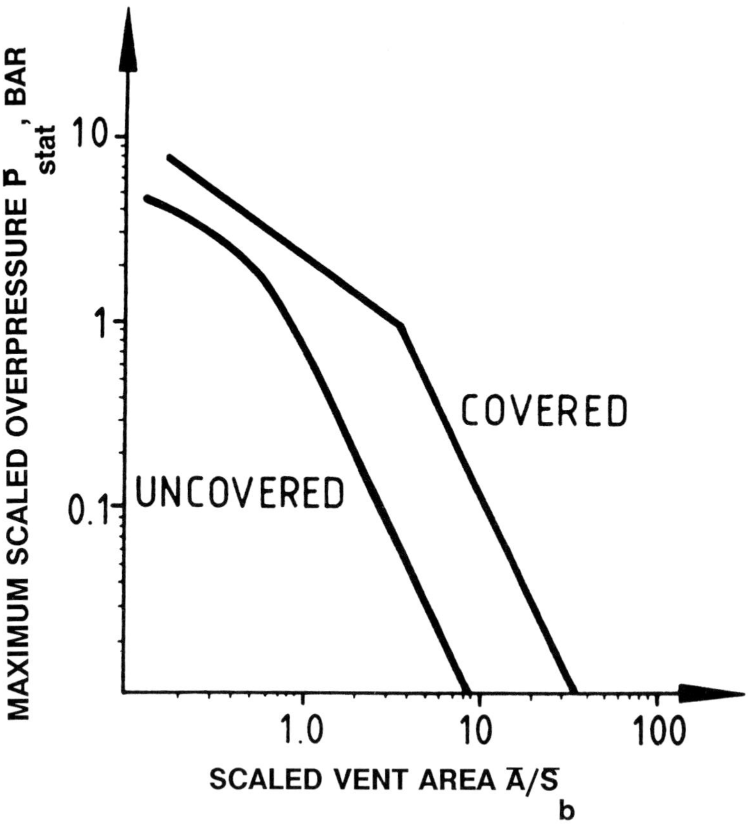

Fig. 3-4. Vent sizes recommended by Bradley and Mitcheson (1978a & b).

TABLE 3-9. Properties of Selected Gas-Air Mixtures at Initial Conditions of 1 Atmosphere and 298°K [Bradley and Mitcheson (1978b)]

Gas	Mole % in Air	ϕ	S_o m sec^{-1}	ρ_{uo}/ρ_{bo}	p_e Atmospheres	$\bar{S}_o$
CH_4	9.48	1.00	0.43	7.52	8.83	8.5×10^{-3}
C_2H_2	7.75	1.00	1.44	8.41	9.78	3.2×10^{-2}
C_2H_2	9.17	1.20	1.54	8.80	10.28	3.7×10^{-2}
C_2H_4	6.53	1.00	0.68	8.06	9.39	1.4×10^{-2}
C_3H_8	4.02	1.00	0.45	7.98	9.31	9.6×10^{-3}
C_3H_8	4.30	1.07	0.46	8.09	9.48	9.9×10^{-3}
C_3H_8	5.00	1.26	0.38	7.97	9.55	9.2×10^{-3}
C_3H_8	6.00	1.52	0.15	7.65	9.30	3.0×10^{-3}
C_5H_{12}	2.55	1.00	0.43	8.07	9.42	9.0×10^{-3}
C_5H_{12}	2.70	1.06	0.43	8.18	9.76	9.3×10^{-3}
C_5H_{12}	3.00	1.18	0.40	8.16	9.77	8.4×10^{-3}
C_5H_{12}	3.50	1.39	0.29	7.92	9.80	6.3×10^{-3}
$C_{16}H_{34}$			0.39	7.82	9.40	7.1×10^{-3}
H_2	29.50	1.00	2.70	6.89	8.04	4.4×10^{-2}
H_2	40.00	1.60	3.45	6.50	7.78	5.3×10^{-2}
TOWN GAS	25.00	1.40	1.22	6.64	8.03	1.9×10^{-2}

$$\phi = \text{Equivalence ratio} = \frac{\text{Actual Fuel / Air Volume Ratio}}{\text{Stoichiometric Fuel / Air Volume Ratio}}$$

p_e = Theoretical closed vessel maximum explosion pressure

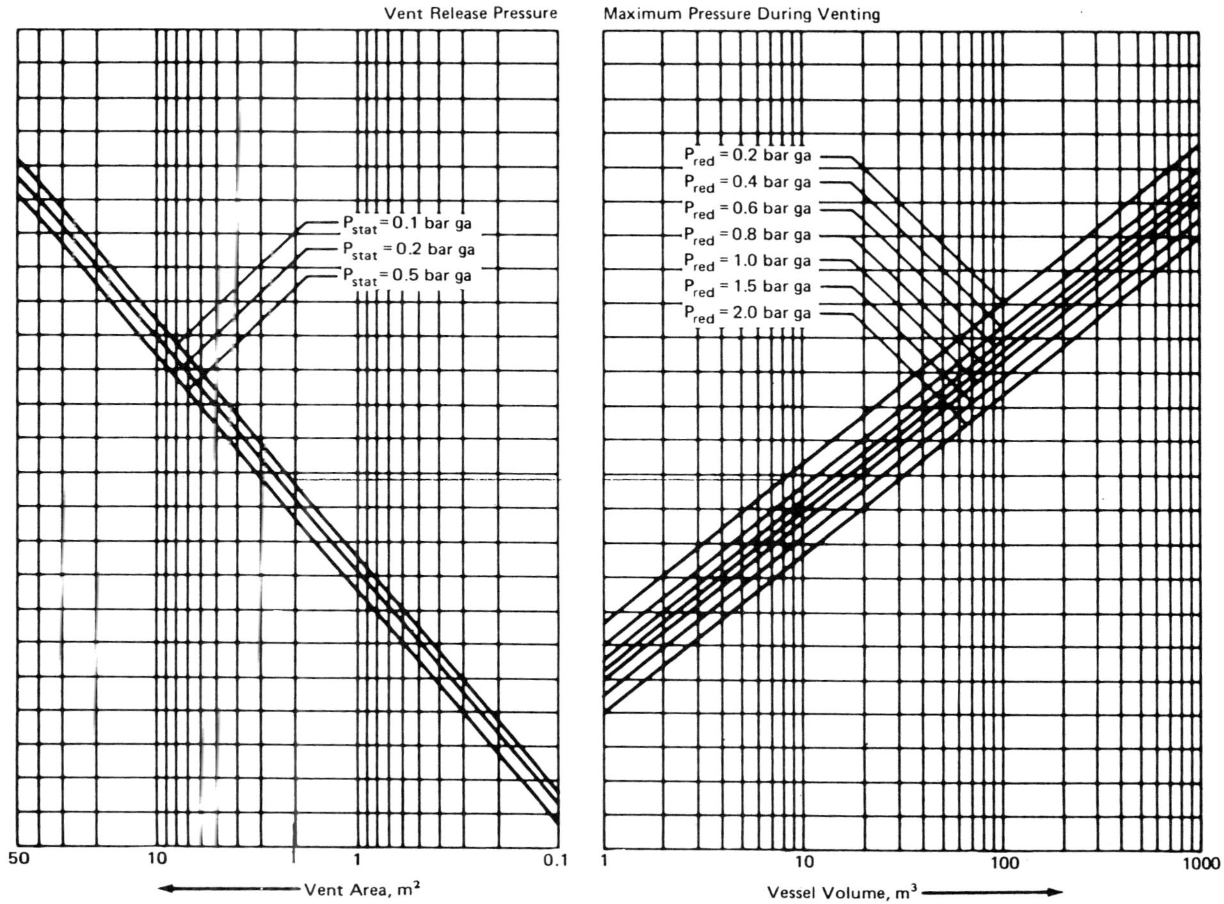

Fig. 3-5. Venting romograph for hydrogen [Bartknecht (1981), NFPA68 (1988)].

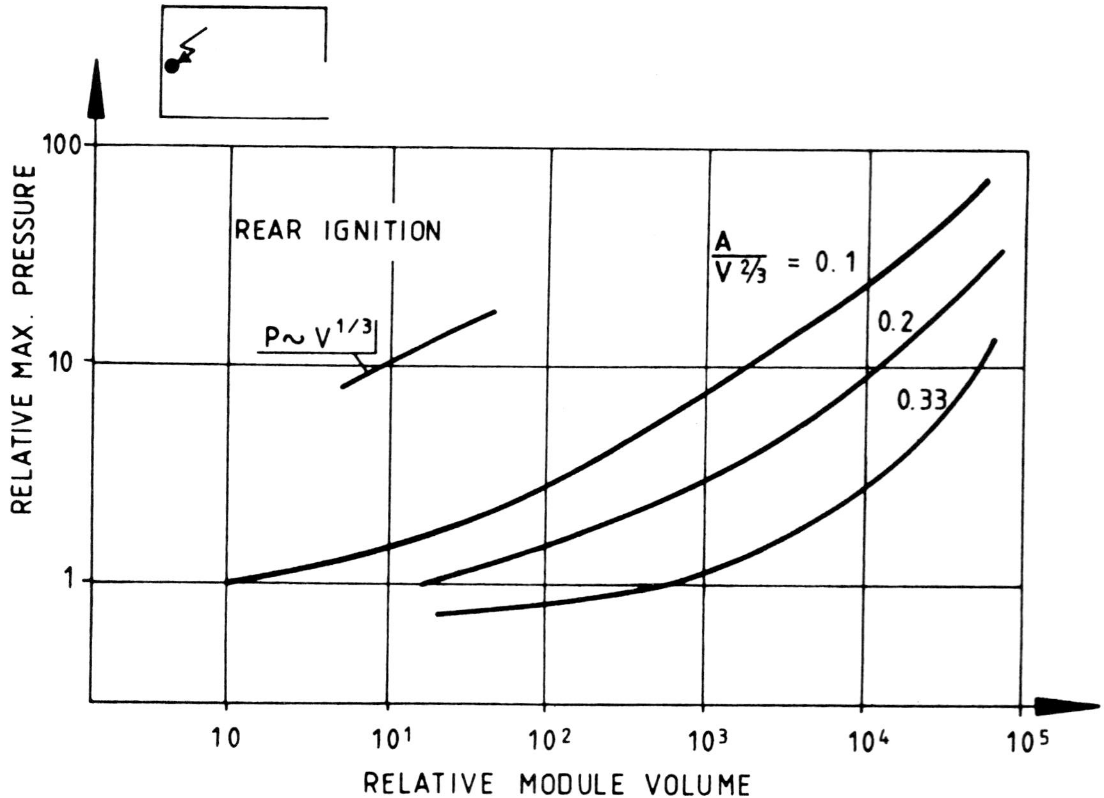

Fig. 3-6. Relative increase in maximum pressure as function of relative increase in compartment volume based on results from Det Norske Veritas experiments [Pappas (1984)].

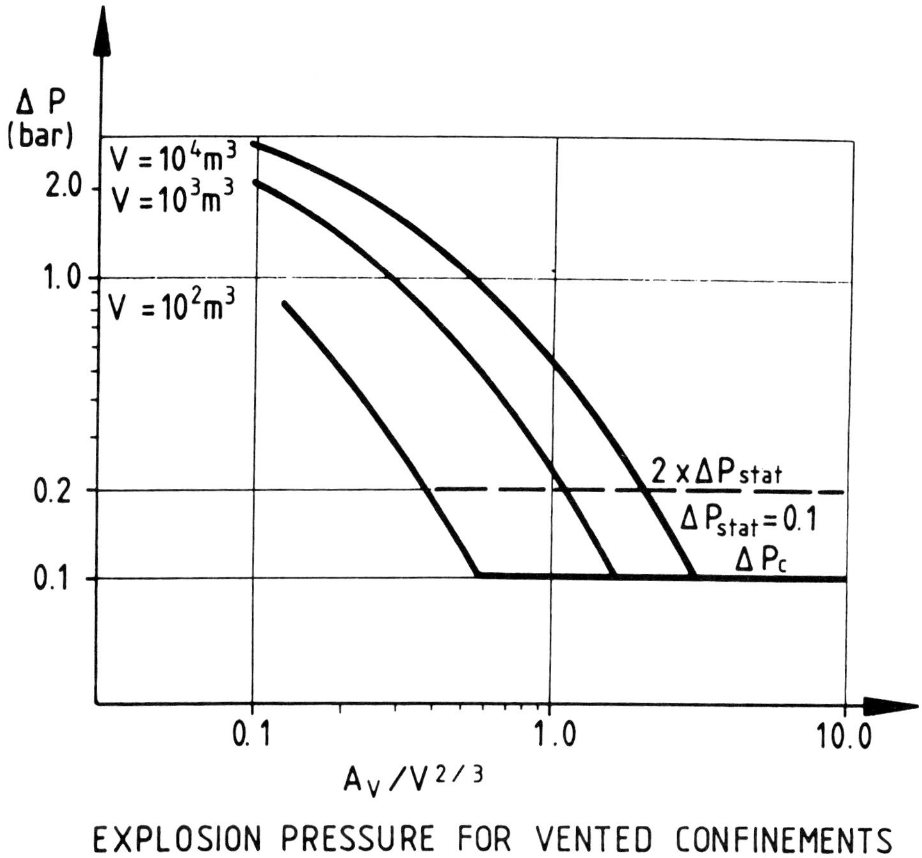

Fig. 3-7. Suggested Veritas relations for estimating necessary vent areas as function of compartment volume [Pappas (1984)].

overall explosion phenomena. Also of importance are the limit conditions within which an explosive mixture can sustain a detonation wave. There is, unfortunately, still no quantitative theory to predict whether a given reactive gas can detonate, or whether the detonation can be initiated by a given ignition source. Nevertheless, some empirical relations and available data will be given in the later discussions.

1. Determination of detonation parameters

The well known ZND model was proposed by Zeldovich, Doering and Von Neumann independently. It assumes one-dimensional flow with the shock at the head of the wave being a jump discontinuity in which no chemical reaction takes place. Following the shock is the reaction zone in which chemical reaction proceeds and is complete in the final state, which is assumed in thermochemical equilibrium. The flow in the reaction zone is steady in the shock-fixed coordinate system. Hence, the shock and the reaction zone, which is usually referred to as the detonation front, propagate together at a constant speed. The propagation speed of the detonation front is usually called detonation velocity and is denoted by "D".

The simplest detonation theory, the so-called Chapman-Jouguet classical theory, is the limit of instantaneous reaction of the ZND model, with the reaction zone collapsed into a single jump discontinuity in which the combustion reaction is completed at infinite reaction rate. The C-J theory, although formulated by Chapman and Jouguet independently nearly a century ago, still has been widely used to predict the final state of detonation products and detonation velocity. This is due to its simplicity and the remarkable agreement between the calculated results and experimental data.

The C-J theory contains a set of equations including mass, momentum and energy conservation equations across the wave front, the equation of state of a perfect gas and the C-J criterion. The C-J criterion is that the detonation wave speed is exactly equal to the speed of the head of the expansion wave following the detonation front, i.e., the sonic condition D=u+c, where u and c denote the flow velocity and local sound speed immediately after the detonation front. Then the five unknown parameters immedicately after the detonation front, namely, p_1, ρ_1, T_1, U_1, and D, can be solved by the five equations as the function of initial conditions in unreacted gas. It is not necessary at all to consider the details of the reaction rate processes inside the reaction zone. Simple equilibrium thermodynamic computations will give the final state of the gas and detonation velocity with satisfactory accuracy even for near limit compositions.

The conservation equations and complementary equations of the C-J theory have been presented in many books and hence are not described here to avoid duplication.

Standard computer codes [Gordon and McBride (1976), Cowperthwaite and Zwister (1973)] are available for the computation of C-J detonation parameters for any fuel-oxidizer mixture at any initial conditions, providing that the enthalpy of formation and other thermodynamic properties of the gases are then in JANAF table or other sources. Detonation parameters for some gaseous mixtures calculated by the Gordon-McBride code are presented in Table 3-10.

Despite the complex multi-dimensional cellular structure of a real detonation wave, the prediction of detonation parameters obtained from the one dimensional C-J theory agrees well with experimental measurements.

The experimental measurement of detonation parameters in a gaseous system is quite straightforward and hence will not be discussed here. Among them, detonation velocity is the characteristic of the system which can easily be measured with high accuracy and hence is often used to verify the correctness of theoretical models.

The most important point is that knowledge of the propagation mechanisms of the detonation wave, such as rate of chemical reactions, detailed structure of the wave front, etc., is <u>not</u> required in determining the steady detonation parameters. In other words, they are uniquely determined once initial thermodynamic conditions are given. This is different from a flame where thermodynamic and gasdynamic analysis do not predict a unique flame speed for a given mixture. This is because the detonation wave propagates with a supersonic speed into the unreacted mixture which has not been disturbed by the wave propagation, whereas the unburned mixture ahead of a deflagration wave has been influenced by the expansion of the combustion products.

2. Direct initiation of a detonation wave

For most practical purposes the initiation problem is much more important than that of detonation parameters. As can be seen from Table 3-10, there is very little spread in the values of the detonation parameters of different fuels. Therefore, the blast effects are not much different from various gases or vapors if they are detonated. However, whether a reactive gas detonates or not makes substantial difference in the blast effects. So, initiation plays a central role in the overall explosion phenomena. Furthermore, once a reactive gas is detonated, we have shown that the detonation states can be predicted quite accurately by the classical C-J theory. This is not the case for the initiation problem, unfortunately, because there is no quantitative theory to the present time to predict whether a detonation can be initiated by a given ignition source.

There are, in general, two modes of detonation initiation, namely the direct initiation of detonation and the transition from deflagration to detonation (DDT). The direct initiation is the fast mode where a detonation wave is initiated in the immediate vicinity of a strong ignition source (sometimes called

TABLE 3-10. Detonation Properties for Selected Stoichiometric Fuel-Air Mixtures*

Fuel	Formula	% Fuel (vol. %)	Chapman Jouguet		
			Pressure (10^5 Pa)	Temperature (K)	Velocity (m/s)
Methane	CH_4	9.50	17.19	2781	1804
Ethane	C_2H_6	5.66	17.98	2815	1803
Propane	C_3H_8	4.03	18.06	2822	1800
Butane	C_4H_{10}	3.13	18.11	2825	1798
Heptane	C_7H_{16}	1.87	17.77	2796	1784
Decane	$C_{10}H_{22}$	0.91	17.76	2794	1782
Ethylene	C_2H_4	6.54	18.36	2926	1825
Propylene	C_3H_6	4.46	18.30	2889	1811
Butylene	C_4H_8	3.38	18.30	2877	1806
Acetylene	C_2H_2	7.75	19.12	3112	1867
Propyne	C_3H_4	4.99	18.78	2999	1834
Butine	C_4H_6	3.68	18.76	2967	1828
Butadiene	C_4H_6	3.68	18.44	2928	1812
Hydrogen	H_2	29.59	15.84	2951	1968
Benzene	C_6H_6	2.72	17.41	2840	1766
Alcohol	C_2H_5OH	6.54	17.68	2735	1773
Ether	$C_4H_{10}O$	3.38	18.26	2819	1796
Propylene Oxide	C_3H_6O	4.99	18.43	2886	1810
MAPP	$C_3H_{5.08}$	4.69	18.34	2921	1818

* Calculated by using Gordon-McBride code at $P_o = 1.013 \times 10^5$Pa
$T_o = 298$ k

"hard ignition"). The DDT is the slow mode of initiation in which a deflagration is first initiated by a weak ignition source (sometimes called "soft ignition") and the detonation is realized through various flame acceleration mechanisms. A comprehensive review of the two modes of detonation initiation has been given by Lee (1977).

As regard to the fast mode of direct initiation, a general description of the mechanism is first presented, followed by a discussion on source characteristics.

1) The mechanism of the direct initiation of detonation

By far the most important means of direct initiation of a detonation is via the strong blast wave generated by a powerful energy source such as a capacitance spark, a laser spark, an exploding wire, or a charge of high explosive. The mechanism of the blast initiation is then via the shock compression of the reactive medium to achieve autoignition temperature corresponding to the shock conditions, which is in the order of 1000-1500K for most hydrocarbon fuels, to induce chemical reactions in the shocked medium. The energy released by chemical reactions then supports the shock and prevents it from decaying . When the energy released by chemical reactions in the shocked medium can fully support the shock, a stable or so-called Chapman-Jouguet (C-J) detonation wave is established.

Lee (1977) described in detail the propagation regimes of a spherical detonation in an acetylene-oxygen mixture. When the initiation energy exceeded a critical value, the shock front and reaction zone were always coupled and a detonation wave propagated into the unreacted gas with, at first, above C-J velocity and then approaching gradually the C-J velocity of the mixture. This is referred to as the supercritical regime. On the other hand, when the initiation energy was below the critical value, the shock wave generated by the initiation source decoupled progressively from the reaction front due to the low shock strength which induced chemical reactions in the shocked medium with ignition delay too long to couple with the shock wave. Thus, the shock decayed and the subsequent propagation of the reaction front was identical to an ordinary flame. This has been referred to as the subcritical regime. When the initiation energy was at the critical value, coupling and decoupling of the shock and reaction fronts occurred interchangeably and the oscillation of the parameters was observed during the detonation propagation.

2) Source characteristics

The most important parameter that characterizes the initiation process is the threshold of initiation energy, so called critical initiation energy. This energy requirement is, of course, dependent on the sensitivities of fuel-oxidizer mixtures. In fuel-oxygen systems, the required energy is of the order of joules and electrical sparks or exploding wires can be used to achieve direct initiation. However, for insensitive fuel-air mixtures, the critical energy is of the order of megajoules and charges of high explosives must be used to achieve direct initiation of detonation in spherical geometry.

Meanwhile, the energy deposit characteristics of the source have substantial effects on the critical value of the initiation energy. It was found that using the same type of igniter, the critical energy for the same mixture can differ by three orders of magnitude when the energy deposit characteristics varied [Bach, Knystautas and Lee (1970)]. A systematic study of the energy-time characteristics of the electric spark as the ignition source has been carried out by Knystautas, et al (1974, 1976) and it was revealed that only the energy released up to the first quarter-cycle of the ringing damped discharge was relevant to initiation. Dabora (1982) also demonstrated the energy-power relationship in a stoichiometric hydrogen-oxygen detonation initiated by a detonator tube and defined the minimum requirement for both initiation energy and power. The above experimental results indicated that the ignition source must be capable of generating a shock wave of a certain minimum strength and the shock wave must be maintained at or above this minimum strength for a certain minimum duration. In other words, the incident shock wave must be of enough strength to compress the fresh medium to a high enough temperature and maintain this temperature for a time duration at least equal to the induction time corresponding to this temperature. Based on the above experimental results Abouseif and Toong (1982) proposed a simple theoretical model to determine the power-energy relation and predict their respective threshold values by means of a constant velocity piston model. This model was later extended by Kailasanath and Oran (1983) to determine the relation between the energy and power required for the initiation of planar, cylindrical and spherical detonations in gaseous mixtures. However, difficulty always lies on the lack of available data of the induction time in various conditions which have significant effect on energy-power relations.

When charges of high explosives are used as the ignition source, the blast wave generated can be considered as an ideal point spherical blast characterized by the total energy only, since the time of energy disposition is short compared with the characteristic time of the blast wave [i.e., R_o/C_o where $R_o = (E_o/P_o)^{1/3}$ is the explosion length and C_o is the sound speed in the medium, according to Lee (1984)]. Nevertheless, there are still problems on how much of the total energy is used. In other words only a portion of the total energy actually goes into initiation and to determine exactly the fraction still remains a problem.

Ohyagi, et al (1985) measured the energy requirement for methane-oxygen mixtures in a planar shock tube and found that the initiation energy estimated by the blast decay method was about 26 to 42% of the total energy released by the initiator tube. A similar factor obtained by Tang, et al (1986) for the direct initiation of a decane spray in a planar tube is only around 15% which is decreasing with the increase of the initiation blast Mach Number. It is concluded that because of the energy dissipation in the mechanical vibration and distortion of the shock tube system, the heat loss and the energy remained in the combustion products of the initiator gas, only a small portion of the total energy released by the initiation source is actually transferred to the tested mixture.

The equivalence of the explosion length R_o in the different geometries of initiation supported by the experiments of Ramamurthi (1976) enables the extrapolation of critical initiation energy from different geometries. In other words, if the critical energy per unit area for a given mixture for the case of planar initiation is known, the corresponding critical energy for spherical initiation can be found since the explosion length is the same for all geometries, i.e., $E_{spherical}/E_{cylindrical} = E_{cylindrical}/E_{planar} = R_o$, where $R_o = (E_o/P_o)^{1/(j+1)}$, j = 0, 1, 2 for planar, cylindrical and spherical geometries, respectively, E_o is the source energy and P_o is the initial gas pressure. The experimental data that support the above conclusion are shown in Table 3-11.

TABLE 3-11. Value of R_o in the Different Geometries of Initiation [Lee (1977)]

Gas Mixture	R_o with Spherical Initiation (cm)	R_o with Cylindrical Initiation (cm)	R_o with Planar Initiation (cm)
100 torr $2C_2H_2 + 5O_2$	2.8	3.0	--
200 torr $2C_2H_2 + 5O_2$	1.23	1.3	--
300 torr $2C_2H_2 + 5O_2$	0.83	0.82	--
40 torr $2C_2H_2 + 5O_2$	--	7.2	7.5
80 torr $2C_2H_2 + 5O_2 + 6Ar$	7.0	7.4	--
380 torr $2H_2 + O_2$	6.4*	7.4	--
800 torr $2H_2 + O_2$	7.2	6.9	7.5

* Deduced on the basis of detonation diffraction experiments.

It should be noticed, however, the extrapolation of the critical energy in spherical initiation from planar geometry is often inadequate due to the effect of boundary conditions. Therefore, the evaluation of the detonation sensitivity of a given mixture is, in general, based on the experiments conducted in unconfined spherical geometry which will be discussed later in Section F.

3. Detonation structure and dynamic parameters

Although the ZND model has provided an explanation of many detonation phenomena, it fails to describe the mechanism of detonation initiation and propagation. According to Lee (1984), a real detonation front proceeds in a cyclic manner and chemical reactions are essentially complete within a cycle or a cell length. The shock velocity is not constant but fluctuates within a detonation cell about the equilibrium C-J value. In other words, in a detonation wave, the chemical heat release is neither instantaneous as stated by the C-J concept, nor spacially homogeneous behind a shock front as illustrated by the ZND model, but rather is controlled by a system of colliding transverse waves which are generated and sustained by the local explosions [Strehlow (1968)].

The cell width λ is the most important parameter characterizing the structure of the detonation front. It has been considered as the length scale of the global rate of chemical reactions in the detonation front and hence of fundamental significance. The parameters relevant to detonation initiation, such as critical initiation energy, critical tube diameter and detonability limits, and the influence of wall losses all do depend directly on detonation structure, i.e., on the value of cell width λ. These are defined as dynamic detonation parameters by Lee (1984) to distinguish from "static" detonation parameters obtained by C-J theory which is independent of detonation structure.

There is, unfortunately, still no quantitative theory for estimating the value of cell size. Experimental measurement by means of the smoked foil records has been the only successful means to obtain the cell size data so far, although there are serious problems with this method since the fish-scale patterns are highly irregular and the selection of the "correct" cell size requires an experienced eye. Early measurements of the cell size were carried out mostly in low pressure fuel-oxygen mixtures diluted with inert gases [Strehlow & Engel (1969)]. The measurement of cell size in fuel-air mixtures at atmospheric conditions is more important for most practical purposes. Some cell size measurements in stoichiometric fuel-air mixtures at atmospheric pressure were made by Bull, et al (1982) more recently, and a systematic measurement of the cell size in common fuel with air mixtures over the wide range of fuel composition at atmospheric conditions was carried out by Knystautas, et al (1983). The results are summarized in Fig. 3-8.

The solid lines in Fig. 3-8 represent the linear relationship $\lambda = Al$, where l is induction zone length and computed from detailed kinetics [Westbrook (1982)]. By matching with one experimental point, the proportionality constant A can be obtained which differs for different mixtures (e.g., A = 10·14 for C_2H_4, A = 52·23 for H_2). The agreement with experimental data is, in general, within one order of magnitude.

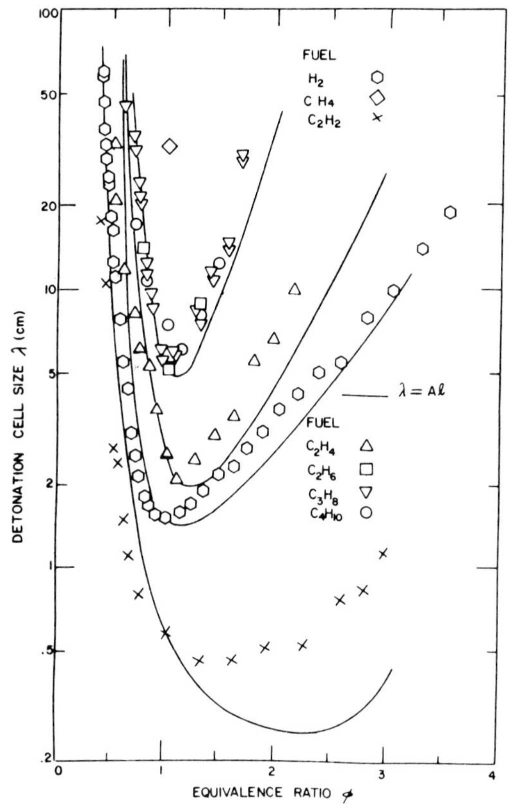

Fig. 3-8. Cell size of fuel-air mixtures at atmospheric pressure [Lee (1984)].

The linear relationship between cell size and induction zone length in part explains why the ZND model, even though physically incorrect, has often provided a good explanation for observed detonation phenomena [Sichel (1989)].

It was found experimentally that when a planar detonation wave propagating in a circular tube emerges suddenly into a large volume containing the same mixture, the planar detonation wave may transform into a spherical detonation wave or fail to propagate depending on the tube diameter. The threshold of the tube diameter d_c is thus defined as <u>critical tube diameter</u>. An empirical relation of $d_c \simeq 13\lambda$ was first observed by Mitrofanov and Soloukin (1965) and then demonstrated by Moen, et al (1981) and Knystautas, et al (1981, 1983).

The relation of <u>critical initiation energy</u> and chemical induction length was first proposed by Zeldovich (1956) and the cubic dependence of the critical energy E on the induction time τ derived from Zeldovich criteria for spherical detonations has been adapted by a number of authors. Based on this criterion and using strong blast theory Lee (1977) derived the expression for the critical initiation energy E_o as the function of chemical induction time τ_c, detonation velocity D and initial density ρ_o of the mixture corresponding to the planar, cylindrical and spherical geometries. However, the lack of the knowledge of the chemical induction time in the detonation wave prevents the above expression and other theoretical efforts from being completely quantitative.

Based on the surface energy concept, i.e., the minimum surface energy with which a planar detonation wave can evolve into a spherical wave without failure, should equal the surface energy of the blast sphere when the initiation blast wave has decayed to the C-J strength. Then equating directly the surface area of the blast sphere to the cross section of the critical tube, $4\pi R_c^2 = \pi d_c^2/4$, the blast energy as a function of the blast radius can easily be determined by Lee, et al (1982) as

$$
\begin{aligned}
E_o &= 4\pi\gamma P_o M_{CJ}^2 I (13\lambda/4)^3 \\
&= (2197/16)\pi\rho_o I D^2 \lambda^3 \qquad (32)
\end{aligned}
$$

where I is a numerical constant that is a function of the specific heat ratio γ (γ = 1.4, I = 0.423 for spherical geometry); ρ_o is the initial density and D is the C-J velocity of the mixture.

Using the cell-size data of Knystautas, et al (1982, 1983), the critical initiation energies have been computed for common fuel-air mixtures and compared with the experimental data obtained by Elsworth from direct experimental measurements of the critical charge. The agreement is reasonably good and Lee's equation can be used to predict critical initiation energies for most practical situations with acceptable accuracy. In this prediction, of course, the availability and accuracy of the cell-size data play a central role.

Detonability limits, in general, refer to the composition limits within which an explosive mixture can sustain a detonation wave. Beyond the lean limit, the mixture contains too little fuel while beyond the rich limit the mixture contains too much fuel to sustain a detonation. It is obvious that the detonation limit is one of the most important parameters in evaluating the explosion hazards of a given fuel and determining the safety conditions whereby a given operation can be performed.

Like the other dynamic detonation parameters, the detonability limit cannot be determined from theoretical considerations alone, but has to be measured experimentally. The U-shaped curves of critical initiation energy versus fuel equivalence ratio can be used to define the detonability limits, however, by specifying somewhat arbitrarily an upper limit of the critical energy above which the mixture is rendered nondetonable. Because of the difficulty in the experimental measurement of detonability limit (high initiation energy requirement and large scale field test for fuel-air mixtures) only fragmental data are available so far. These data are given in Table 3-12 along with the references. It should be noted that these data are very approximate and apparatus dependent. Therefore, interested readers should refer to the references for details.

It can be seen from the table that detonability limits are, in general, narrower than the flammability limits of the same mixture.

Based on the criteria for stable propagation in tubes, detonability limits can also be predicted from a knowledge of the cell size λ. It was found by Donato (1982, 1981) that the onset of the single-head spinning waves in a given tube, which corresponds to a cell size of the order of the tube circumference (i.e., $\lambda \simeq \pi d$), should correspond to the detonability limits for that tube. Thus, the composition limits for a given tube can be estimated if λ is known. In other words, if the onset of single-head spinning waves is found experimentally in a tube, the composition limits can be obtained from the known relation of λ versus composition, using $\lambda = \pi d$ where d is the diameter of the tube.

4. Deflagration to detonation transition (DDT)

A deflagration wave is subsonic and can thus perturb the gas ahead of the flame. In general, compression waves are generated ahead of the flame due to the expansion of combustion products. Since a newly produced compression wave propagates at a higher velocity through the gas which is precompressed and heated by the previous compression waves, a shock wave will be formed and strengthened when overtaken by successive shock or compression waves. When the strength of the precursor shock reaches a critical value, a sudden change of the propagation mechanism occurs from the slow process controlled by diffusion and heat conduction to adiabatic compressive heating. This corresponds to the onset of a detonation which is not a continuous process from deflagration to detonation, but rather the "sudden" formation of the detonation. Just prior to the onset of detonation, the universal picture according to Lee (1977) is one of a highly turbulent flame brush propagating behind a train of intense shock waves and the establishment of detonation is originated from "explosion in the explosion" as Oppenheim called it. The local explosion results from the formation of a "hot spot" due to finite gasdynamic fluctuations in the precompressed unburned gas. Although the gasdynamic fluctuations that trigger off the localized explosion in unburned gases may have numerous causes, the average condition of the unburned gases between the shock and flame brush must be close to the autoignition limit of the mixture which are of the order of 1100K and 1500K for hydrocarbon-oxygen and hydrocarbon-air mixtures, respectively. Therefore, the minimum shock

TABLE 3-12. Detonability of Some Fuel-Air and Fuel-Oxygen Mixtures

	Fuel-Air Mixture		Fuel-Oxygen Mixture	
Fuel	Stoich. Comp. (volume %)	Detonability Limits (volume %)	Stoich. Comp. (volume %)	Detonability Limits (volume %)
Acetylene (C_2H_2)	7.75	6.7 - 21.4[1]		18 - 59[1]
			28.6	6.75 - 68[1]
Hydrogen (H_2)	29.59	13.5 - 70[2]		
		18.0 - 59[3]	66.66	39 - 81[1]
Ethane	5.66	4.34 - 8.64[4]	22.22	12 - 38[1]
		2.87 - 12.2[5]		11 - 39[1]
Propane	4.03	3.0 - 7.0[2]	16.66	8 - 30.5[1]
		2.57 - 7.37[5]		7 - 31[1]
		2.2 - 9.2[1]		
Butane	3.13	2.2 - 5.4[2]	13.33	
		1.98 - 6.18[5]		
Ethylene	6.54	3.3 - 14.7[5]	25.0	15 - 48[1]
				9.75 - 51[1]
Propylene	4.46	3.5 - 8.5[2]	18.18	9 - 35[1]
		3.55 - 10.4[5]		6.75 - 37[1]
Ethylene Oxide	7.75	5.3 - 18.0[1]	28.57	13.5 - 60.5[1]
				8.5 - 65[1]
MAPP*	4.69	2.9 - 10.2[2]		
		2.9 - 10.5[6]		
		3.3 - 9.8[7]		

[1] Lee & Matsui (1978)
[2] Benedick (1983)
[3] Lewis & Von Elbe (1961)
[4] Bull, et al. (1982)
[5] Borisov & Loban
[6] Fry & Nicholls (1974)
[7] Collins (1974)

* MAPP = 51.0% Methyl acetylene, 26.0% propane, 23.0% propadiene

strength required is $M_s \simeq 4$ and $M_s \simeq 5$ for fuel-oxygen and fuel-air mixtures, respectively.

The strength of the precursor shock depends on the flame speed, the ratio of the specific volume increase across the deflagration wave and the boundary conditions. The ratio of the specific volume increase is nearly a constant of about 7 for common hydrocarbons, hence the occurrence of the transition depends only on the flame speed corresponding to the boundary conditions. The flame speed increases drastically via various mechanisms of flame acceleration. So, the DDT problem is actually a problem of flame acceleration.

According to an analytical model proposed by Lee, et al (1976), which is based on the assumption that the critical states are achieved via the adiabatic compression of a single shock wave generated by the turbulent flame, the criterion of the turbulent flame speed required for the transition to occur is about 230 m/sec for fuel-air mixtures and about 65 m/sec for fuel-oxygen mixtures. The laminar flame speed for mass hydrocarbon-air mixtures is of the order of 0.5 m/sec, thus an amplification factor $u_t/u_l \simeq 460$ is required to produce the critical states for the DDT to occur. On the other hand, the laminar flame speed is about 10 m/sec for fuel-oxygen mixtures and hence the amplification factor required is only $u_t/u_l \simeq 6$. This explains the great ease of the establishment of a detonation in fuel-oxygen mixtures, as compared to fuel-air mixtures. The fact that for mixtures of acetylene and hydrogen with air, the laminar flame speed is of the order of 1.5 m/sec, which is about three times higher than most hydrocarbon fuels, also explains the relative ease of the realization of DDT in these particular mixtures, since the amplification required is much less than for most hydrocarbon-air mixtures.

By far the most important mechanism of flame acceleration is that due to turbulence created by obstacles in the path of the propagating flame. Consider a gas mixture ignited by a weak source. A laminar flame propagates in the quiescent reactive gas. Due to the expansion of the burnt gas, the unburnt mixture ahead of the flame will be set into motion with a velocity $u = (\Delta v/v)s$, where u is the velocity of unburned gas relative to laboratory coordinate, s is the velocity of burned gas relative to the unburned gas, and $\Delta v/v$ is the ratio of specific volume increase across the flame to the initial specific volume. For most hydrocarbon-air mixtures with a typical laminar burning velocity of about 0.5 m/sec and the increase in specific volume across the flame of 7 times the initial specific volume, the flow velocity of unburnt gas will be of the order of 3.5 m/sec. Therefore, no significant flame acceleration is observed in a completely unconfined spherical flame. However, if solid surfaces of the confinement or obstacles are present in the flow field, shear layers and velocity gradients will be formed in the boundary layers on the surfaces as well as in the wake of the obstacles. As the flame advances into the velocity gradient field, the flame surface will be stretched and folded. The distorted flame will consume unburnt gas over a larger surface area leading to an increase in the rate of heat release which corresponds to a higher volumetric or mass burning rate. Turbulence in the shear layers also enhances the local burning velocity because of the higher turbulent transport rates. This combined effect of large scale flame folding and fine scale turbulent enhancement of the local burning velocity causes the volumetric burning rate to increase. The higher burning velocity results in a larger displacement flow velocity of the unburned gas ahead of the flame. This in turn gives rise to a more severe velocity gradient and higher turbulent intensity in the shear layers. Thus, the flame speeds of the order of hundreds of meters per second and even DDT is achieved due to the positive feedback loop [Lee (1977)].

Both experimental and numerical studies on the subject of flame acceleration have been carried out extensively during the last decade due to its significance in accidental explosions. A number of reviews have already been written by Lee (1977, 1983) and Moen (1982) describing the important progress made in this field. It has been demonstrated that flames, even in rather insensitive fuel-air mixtures, can, under appropriate conditions, accelerate very rapidly to very high speeds and develop very high overpressures. The acceleration rate, the flame speed attained and the associated overpressures developed are all functions of the particular fuel, mixture composition, obstacle configuration, confinement and geometry of the flame. Recent experimental results of the dependence of the turbulent flame speed on fuels, compositions and obstacle configurations are summarized by Lee (1984).

It should be noticed that for the onset of detonation, the critical states must be maintained for a certain minimum duration which should be at least of the order of magnitude of the induction time corresponding to the autoignition temperature. For fuel-oxygen mixtures, the induction time is of the order of microseconds while that for fuel-air mixtures is one to two orders of magnitude longer. This explains why in some cases when the high turbulent flame speeds were produced artificially, the transition process still could not be realized because the folded structure of the flame could not be maintained long enough for the onset of detonation.

F. Unconfined vapor cloud explosions (UVCE)

1. Introduction

Experimental and analytical studies on UVCEs have been pursued by two groups of investigators with totally different purposes. One group is focused on the development of the weaponry system termed FAE for Fuel Air Explosives, and the other group is concerned with avoiding severe UVCE accidents. The fuels in FAE weapons are usually sensitive to shock initiation, e.g., ethylene oxide, propylene oxide or MAPP gas (a mixture of methyl-acetylene, propylene and propadiene), the dispersion of fuel and the ignition of combustible mixture are designed to detonate most of the fuel after rapid mixing with air. By contrast, fuels in accidental UVCEs can be any gaseous or flash-evaporating liquid fuels. Only a portion of fuel is within flammable limits, and even less fuel will serve as the source of generating significant blast effects in accidental UVCEs. The estimation of explosion yield is based on only 2 to 10% of the heat of combustion of the total fuel, since the explosion yield ranges from 0.1 to 10% with the majority less than 1 to 2% according to accident history analysis [Lind, et al (1982), Davenport (1977)]. The blast yield of Flixborough disaster was estimated to be 5% of the combustion energy from 30 tons of cyclohexane [Sadee, et al (1977)]. The highest yield factor of 10% was reported by Baker (1984).

The formation of an explosive vapor cloud in FAE weaponry system is due to the intentional dispersion of combustible gas or mist in the atmosphere by a well-designed explosive charge, while in an accident scenario, the formation of the explosive vapor cloud is most likely the consequence of an accidental release of a combustible gas or liquid. The accidental release of gaseous or liquid fuel may not necessarily lead to an explosion. Often, there is only a slow burn or only a dispersion of fuel without ignition.

It has been realized that to have a UVCE, certain conditions must be met. First of all, there must exist a pre-mixed vapor-air and/or mist-air mixture which lies within the explosion limits. In other words, there should be a proper delay time between the escaping of fuel vapor or mist into the atmosphere and the ignition of the cloud to allow the formation of the explosive mixture. Secondly, there must be either a strong ignition source (or sources) or some flame acceleration mechanisms to generate a shock wave of significant strength which can cause damage to the surroundings.

Once the above conditions are met and a UVCE is established, the following effects to the surroundings are, in general, realized:

1) A sizeable fireball,
2) A wide spectrum of air blast effects, ranging from minimal to catastrophic, and
3) Minimal fragmentation.

In later discussions, particular regard will be given to the blast effects since fragmentation effects are negligible and radiation effects can be referred to Baker, et al (1983) and other references.

Unfortunately, a considerable number of accidental UVCEs have been recorded throughout the world over the recent years. Some of the massive explosions have been the predominant cause of the largest losses in the chemical and petrochemical industry. The statistical analysis over the 150 losses occurred in the recent 30 years by Garrison (1989) has shown that the magnitude of the losses has increased in each of the five-year periods and that 46%, 42% and 60% of the losses were initiated by UVCEs within petrochemical plants, plastic/rubber plants and natural gas processing, respectively. Furthermore, the UVCEs resulted in highest fatalities and property losses. For example, the Flixborough works of Nypro (UK) Ltd were virtually demolished by an explosion on June 1, 1974 due to the accidental release of about 30 tons of cyclohexane. This accident killed 28 people and injured many more and the damage to property extended over a wide area. Very recently, a massive explosion on October 23, 1989 destroyed Phillips Petroleum Co. plastics facility 15 miles from Houston, Texas, with 24 killed and 124 injured. The fire and explosion were probably fueled by ethylene and isobutane gases--both used in plastic manufacturing and plastic products.

Because of trends toward plants of larger capacity, higher pressures and temperatures and increasing quantity of shipping and storage of combustible liquids and gases, the losses resulting from UVCEs have been increasing both in frequency and severity. Therefore, there has been a major growth of concern on the subject of unconfined gas and vapor cloud explosions during the recent years. Several surveys of vapor cloud incidents [Strehlow (1972), Strehlow and Baker (1975), Davenport (1977), Gugan (1979)] have summarized the accidental explosions, their damage and available data. Davenport (1977) summarized the type of facility, type of plant and mode of release for incidents where overpressures were created. His statistics have shown that industrial, particularly chemical and petrochemical, plants and the process equipment are the major areas for encountering UVCEs. Selected industrial incidents were listed and amount of loss for each incident were estimated. More complete statistics on vapor cloud incidents were made by Gugan (1979). Incidents involving large clouds that upon ignition did not create noticeable overpressures and incidents where large clouds did not ignite are also included in both surveys.

2. Direct initiation of an unconfined detonation

Like other types of explosions, UVCEs can also be categorized into two types, namely deflagrations and detonations, according to the different propagation mechanisms. Detonation is, of course, the most violent form which can cause the most severe damage to the surroundings over very large distances. Also like other types of explosions, the detonation of an unconfined vapor cloud can be achieved through either the direct initiation or the transition from deflagration. In either case there are quite restrictive conditions which must be satisfied if the detonation is to propagate.

The most important parameter, which characterizes the direct initiation of a combustible mixture, is the critical initiation energy. The critical initiation energy is defined as the minimum of source energy required for the direct initiation of an unconfined detonation in a reactive mixture. It has always been used to illustrate the detonation sensitivity of the given mixture, since it avoids all external influences such as boundary confinements and relates directly to the rates of chemical reactions in the detonation front.

The critical initiation energy data measured directly for unconfined fuel-air clouds are very limited due to the experimental difficulties. The fragmentary data available are briefed below.

1) The direct initiation of methane-air mixtures

In view of the safe transport and storage of liquefied natural gas (LNG), the direct initiation of methane-air mixtures has been studied by several investigators in the 1970 decade. Bull, et al (1977) used stoichiometric methane-oxygen mixtures diluted with various amounts of nitrogen. The gaseous mixtures were contained in polyethylene bags of uninflated size 1.8m x 1.8m and 3.05m x 1.52m and initiated by tetryl charges placed on one end of the bag. They were able to detonate mixtures of $CH_4 + 2O_2 + xN_2$ in the range of $0 \leq x \leq 5.5$. The critical initiation energy for methane-air mixture (i.e., mixture of x = 7.2) could not be obtained directly because of the limitation of the experimental arrangements. Nevertheless, a linear relation was found between the logarithm of tetryl charge weight to nitrogen concentration. The extrapolation of the straight line indicated that 22 kilograms of tetryl would be required to initiate a spherical detonation in a stoichiometric methane-air mixture. The results also pointed to the long path length of eleven meters which would be required in any experimental verification of the extrapolation.

Kogarko, et al (1966) reported initiating a methane-air detonation with a one kilogram charge of TNT, but Bull, et al (1977) revealed that this measurement was so close to the initiation source that the wave propagation was influenced by the initiation source, not suitable to determine whether or not the wave was self-sustained.

Vanta, Foster and Parson (1974) also reported that natural gas-air mixtures were detonated by one kilogram of high explosive in 1.2m square x 6m long enclosures made of plastic film. However, they noted that the detonation front was atypical of other hydrocarbon-air detonations and that the propagation velocity was considerably lower than the calculated C-J detonation velocity.

Benedick (1979) measured the critical initiation energy of methane-air mixture using large polyethylene film enclosures of 2.4m square cross-section and 12m long that were supported by a framework of steel pipe. To minimize the divergence of gaseous detonation from the initiator the high explosives were 1m x 2m rectangular layer of Detasheet supported on plywood. 3.6 to 4.1 kg of high explosives initiated stable detonations that propagated the full length of the enclosure (12m), while the 1.5 kg and 2.8 kg initiators established detonations that failed after propagating about 1 meter. When the mass of the Detasheet layer was 4 kg but the cross-section of the enclosure was only 1.6m square, detonations were established but failed after propagating about 7m along the bag length. It is concluded from the experiments that stable detonations can be initiated in a stoichiometric methane-air mixture by about 4 kg of high explosives which is much less than the extrapolation by Bull, et al (1977) and the prediction by Boni, et al (1977). However, the cross-section of the initiation charge and unconfined gas column must be sufficient to prevent rarefaction waves from quenching the reaction in a region along its axis. So, the critical initiation energy of 4 kg of high explosive charge obtained by Benedick, et al (1979) for methane-air mixture is not for a completely unconfined spherical cloud, but the requisite cross-section of about 2m square is very small relative to the dimensions of clouds resulted from accidental release of large quantity of fuel.

Westbrook and Haselman (1979) predicted the critical initiation energy of 50-100 kg high explosives for unconfined methane-air mixture using a numerical model which combined blast wave decay and chemical kinetics and was validated by means of comparison with Bull's experimental data (at least one data point).

It has also been demonstrated by Bull, et al (1977) that methane is the most stable among all hydrocarbons due to its particular molecular structure. The amount of tetryl required to initiate detonation in methane-air mixture is more than two orders of magnitude greater than that in heavier paraffin hydrocarbons and more than three orders of magnitude greater than that in ethylene-air mixture. This is in agreement with the measurements of detonation cell size or critical tube diameter. The comparison of the critical energies for the mixtures of some hydrocarbon fuels with air is shown in Fig. 3-9.

The decrease of critical initiation charge of methane with the addition of heavier hydrocarbons, typically ethane, has also been illustrated by Bull, et al (1978) as in Fig. 3-10. The strong influence of small ethane concentrations is readily apparent.

It is clear that the requirement for initiation source for most natural gases, which is at least 1 kg of tetryl [Bull, et al (1977)], is much lower than that for pure methane. This trend is also verified by Nicholls, et al (1979) using a sectored shock tube.

No result of the direct initiation of methane-air mixtures that deviate from stoichiometric proportions has been reported so far.

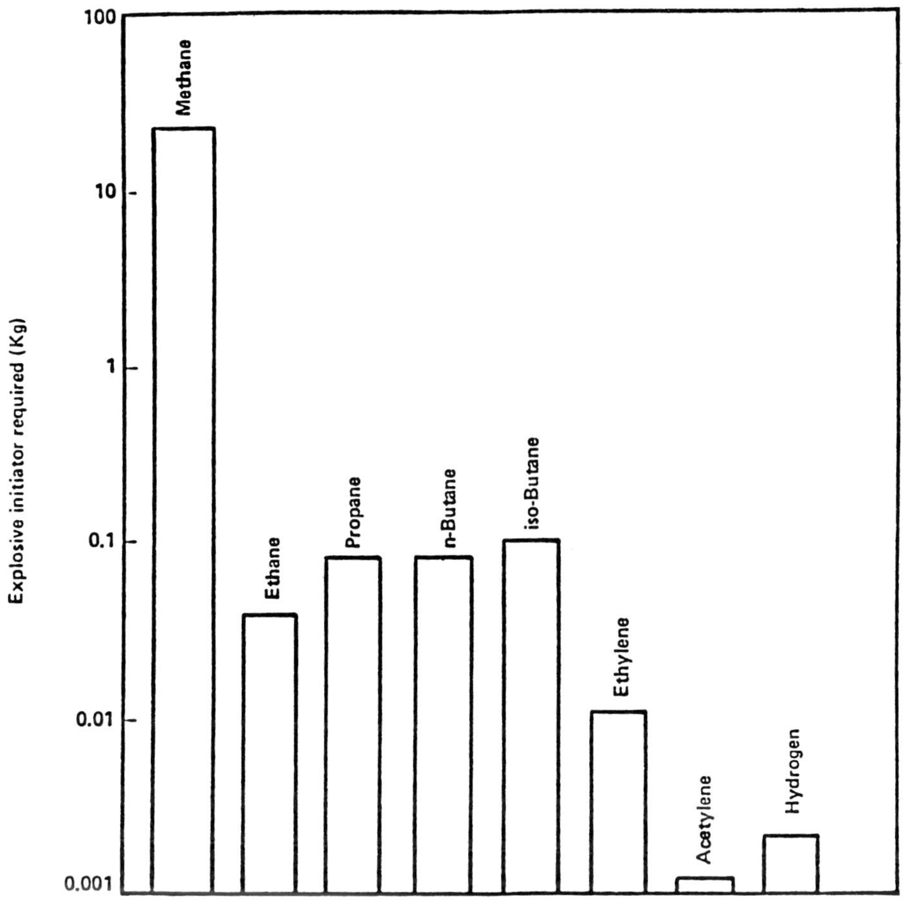

Fig. 3-9. Relative detonabilities of fuel/air mixtures [Bull, et al (1978)].

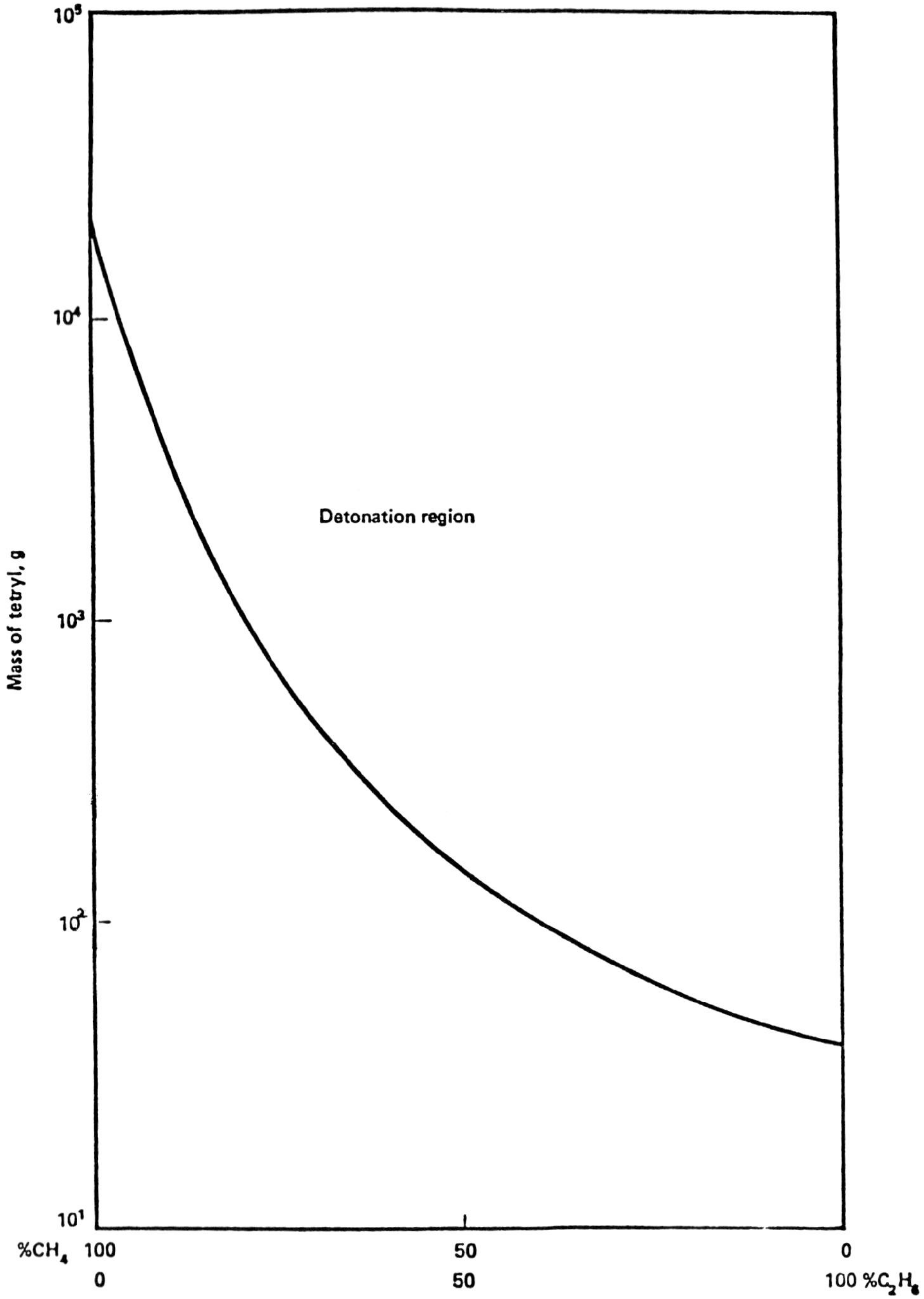

Fig. 3-10. Decrease of critical initiation energy of methane-air mixtures [Bull. et al (1978)].

2) Direct initiation of hydrogen-air detonations

In view of the specific interest in hydrogen-air detonations related to safety operations of nuclear reactors, a series of large-scale experiments were carried out at the field test facilities of Sandia National Laboratories by Benedick, et al (1986) to measure the critical charge weight for direct initiation of unconfined detonation in H_2-air mixtures.

The Sandia experiments were carried out in large, slightly tapered (2.4m to 2.7m in diameter and 4.2m in length), vertically mounted cylindrical polyethylene bags. The initiation charge, composed of two hemispheres of composition C-4 plastic explosive (ΔH_e = 4.87 kJ/gm), was located one meter above the base on the axis of the bag to avoid the influence of ground reflection of the blast wave on the initiation of the upper portion of the bag to ensure completely unconfined conditions.

The results of Sandia experiments are summarized in Table 3-13. The detonation range obtained in Sandia experiments has extended the Elsworth results, which were carried out in an explosion vessel of about 1.5 m^3 in volume, from 17.4% H_2 to 16.7% H_2 on the lean side and 57% to 60% on the rich side. However, for critical initiation energy, the differences between the Elsworth results and the Sandia results are substantial, particularly for lean mixtures. For instance, for a mixture of about 20% H_2, the Sandia charge weight measurement is three-fold of Elsworth's result, while at the composition of about 17.4% H_2, the Sandia measurement is more than twelve-fold larger. It is clear that the Elsworth's results underestimate the required initiation energy, due to the limited size of his apparatus, so that there appears to be insufficient travel for the detonation wave away from the initiation source to assess whether the detonation will fail or not.

For comparison, the recent results of Tiezen, et al (1986) for hydrogen-air detonation obtained in a tube of 0.43m inner diameter, which can be operated at elevated temperatures, are also summarized below.

For H_2-air mixtures at 20°C and a total pressure of 1.0135 x 10^5 Pa, detonations have been achieved between 13.5% and 70% H_2 mole fraction (13.5% and 70% H_2 mixtures correspond to the onset of a single head spin detonation). These limits are much wider than the previously believed 18-59% H_2 by Lewis and von Elbe (1961) obtained in a smaller tube but close to that obtained in McGill by Knystautas, et al (Lee, 1984) in the tube of 30 cm inner diameter. The detonability increases with increasing initial temperature at constant density, e.g., at 100°C and on air density of 41.6 mole/m^3, a 13% H_2-air mixture has been detonated.

The addition of CO_2 or H_2O vapor to H_2-air mixture greatly reduces the detonability of the mixture. The addition of 5, 10 and 15% CO_2 to a stoichiometric H_2-air mixture increases the cell width λ (about 5 mm for

TABLE 3-13. Critical Initiation Energy for H_2-Air Mixtures [Benedick, et al. (1986)]

(p_o = 630 Torr)

% H_2 in Air	ϕ	Charge Weight g (Comp. C-4)	Detonation
16.7	0.48	461.0	No
17.4	0.50	461.7	Yes
18.5	0.54	60.5	No
18.5	0.54	88.5	No
18.5	0.54	151.0	Yes
20.0	0.59	14.5	No
20.0	0.59	30.5	Yes
55.1	2.92	153.0	Yes
57.0	3.15	152.7	No*
58.1	3.30	461.1	Yes
60.0	3.57	461.0	No*
60.5	3.65	152.7	No

* Indirect initation on reflection from boundaries.

stoichiometric H_2-air mixture) by factors of 1.5, 2.8 and 12.8 respectively. According to the cubic power law of Zeldovitch criterion, the addition of CO_2 decreases the detonability by corresponding factors of 3.4, 22 and 2100 respectively. It was also observed that the role of water vapor is to desensitize the mixture by a substantial factor. As an example, the addition of 10, 20 and 30% of steam to the stoichiometric H_2-air mixture leads to the increase in detonation cell size λ by the corresponding factors of 6, 30 and 60. According to Zeldovitch criterion, the data indicate a reduction in detonability of the mixture by factors of 220, 2.7×10^4 and 2.2×10^5, respectively.

3. Blast effects produced by unconfined gas deflagrations

The knowledge of blast effects generated by unconfined gas deflagrations is of importance in accidental explosions. Gugan (1979) reviewed 100 accidents on unconfined vapor cloud explosions from 1921 to 1977 and showed that about 60% of them involved blast effects which resulted in large destructive distances and severe damage. Statistics had also shown that most of the UVCE accidents which produce blast damage are related to deflagration rather than detonation.

It is extremely difficult to evaluate a priori blast effect of a potential explosion hazard, for many reasons. First, the accurate estimation of the amount of fuel released before ignition has often proved impossible, with the best accuracy within a factor of two. Furthermore, only a portion of the fuel could be mixed with air within the explosion limits and the accuracy of the estimation of the amount of fuel premixed with air could be predicted for the order of magnitude. The uncertainty of the portion of fuel which premixed with air but burned in low burning velocity and hence made insignificant contribution to the blast effects also adds additional difficulties.

Despite the above difficulties, several models have been proposed by various investigators to predict the blast effects generated by unconfined vapor cloud deflagrations because of its practical importance in assessing the potential explosion hazards associated with accidental spills of variety of fuels. Some of the models, or the results of the models will be presented at the forms which can be readily used by engineers. However, it is not possible to make a comparison between these models because experimental data on the deflagration generated blast are rare.

1) Spherical flame model by Strehlow, et al (1979)

Strehlow, et al (1979) has carried out a numerical study for blast waves generated by a spherical flame propagating from the center to the edge of the combustible mixture with constant velocity. In the computation, the flame was represented by a heat-addition, "two-gamma" working-fluid model and the real equilibrium Hugoniots corresponding to six commonly used fuel-air mixtures were used to obtain proper values of gamma γ and heat addition Q in this model. The results appear as solid curves in Fig. 3-11 where the scaled overpressure $\bar{P}_s = (P_{max}-P_o)/P_o$, and the scaled radius $\bar{R} = R/R_o$, where $R_o = (E_t/P_o)^{1/3}$; E_t is the total energy deposited in the source region, P_{max} is the maximum of pressure rise observed at $\bar{R}$.

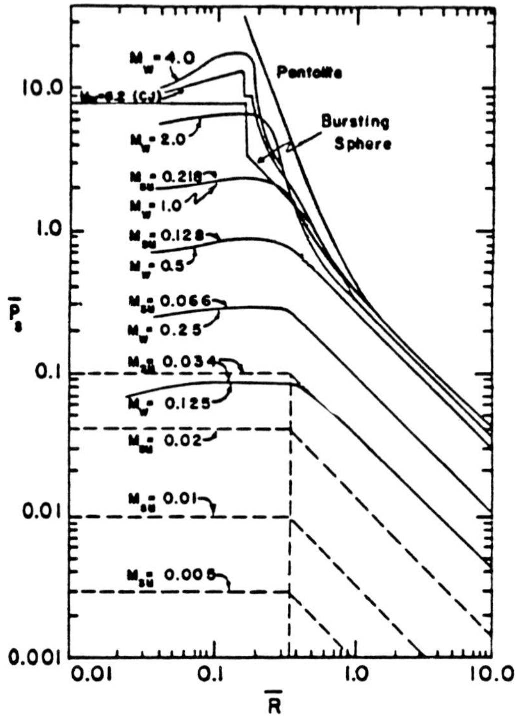

Fig. 3-11. Scaled overpressure vs energy-scaled radius for constant-velocity flames. Solid line: numerical calculations. Dashed lines: Taylor's theory extended to the very low Mach number range [Strehlow, et al (1979)].

Fig. 3-12 presents the scaled positive impulse $\bar{I}_t$ versus scaled distance $\bar{R}$, where

$$\bar{I}_t = \frac{\bar{I}_t a_o}{E_t^{1/3} p_o^{2/3}} . \tag{33}$$

In these figures M_w is the flame Mach number which equals the flame velocity divided by the velocity of sound in the ambient air. M_{su} is the normal burning velocity divided by the ambient sound velocity. These figures also appear in Baker, et al (1983).

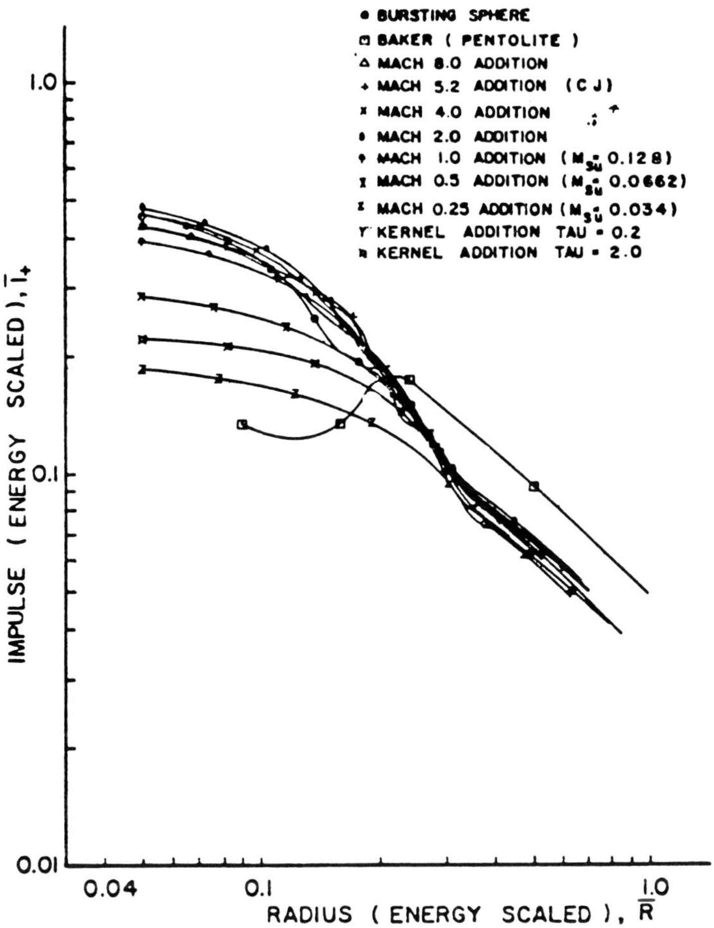

Fig. 3-12. Energy-scaled impulse vs energy-scaled radius for constant velocity flames, a bursting sphere, the ramp addition of energy, and reference Pentolite [Strehlow, et al (1979)].

Numerical calculations were not performed at extremely low flame velocities because computer times become very long and calculation becomes very costly. An analytical expression developed by Taylor (1946) is then used to predict the maximum pressure rise in low velocity cases. The results obtained by using the Taylor's acoustic solution for M_{su} = 0.034 - 0.005 are plotted as dashed lines in Fig. 3-11. The pressure remains constant until it reaches the limit radius of the source region and then begins to drop. The inverse-radius law ($P_s \propto 1/\bar{R}$) is used to extend the curve to larger radii.

Additionally, the effect of flame acceleration on the blast wave was studied and the results show that the maximum overpressure generated by an accelerating flame was always less than that generated by the constant velocity flame traveling at the maximum velocity. Thus the constant velocity results are conservative and can be used to estimate the overpressure, providing that the maximum value of the flame velocity is known. This aspect is discussed in more detail in Baker, et al (1983).

As can be seen from Fig. 3-12, positive specific impulse is essentially unaffected by the flame Mach number. However, the scaled positive impulses produced by the sources of low energy deposit rates are apparently lower than that generated by Pentolite. Nevertheless, the duration of the impulse loads will be considerably longer than that for an equivalent high energy density source.

In summary, the overpressure and positive impulse of the blast wave generated by a spherical deflagration can be readily found from the curves resulted from Strehlow's model. However, flame speed or normal burning velocity should be used as the input and how the flame accelerated to this speed is not addressed in this model.

2) TNO hemisphere model

A shock wave model was presented and a computer program has been developed by the Prins Maurits Laboratory TNO [Pasman (1976), Wiekema (1980)] to allow the estimation of the possible blast effects from the deflagration of a hemispherical combustible cloud. The model is illustrated by Fig. 3-13. A hemispherical cloud with a volume v_o consisting of a homogeneous combustible gaseous mixture will, after ignition in the center, expand to a hemisphere with volume v_1. This expanding movement taking place at various speeds is replaced by an equivalent piston movement. The radius r_1 corresponding to volume v_1 and the time t_1 in which the expansion process is completed gives the average flame velocity as

$$\overline{V}_{J1} = (r_1 - r_o)/t_1 \tag{34}$$

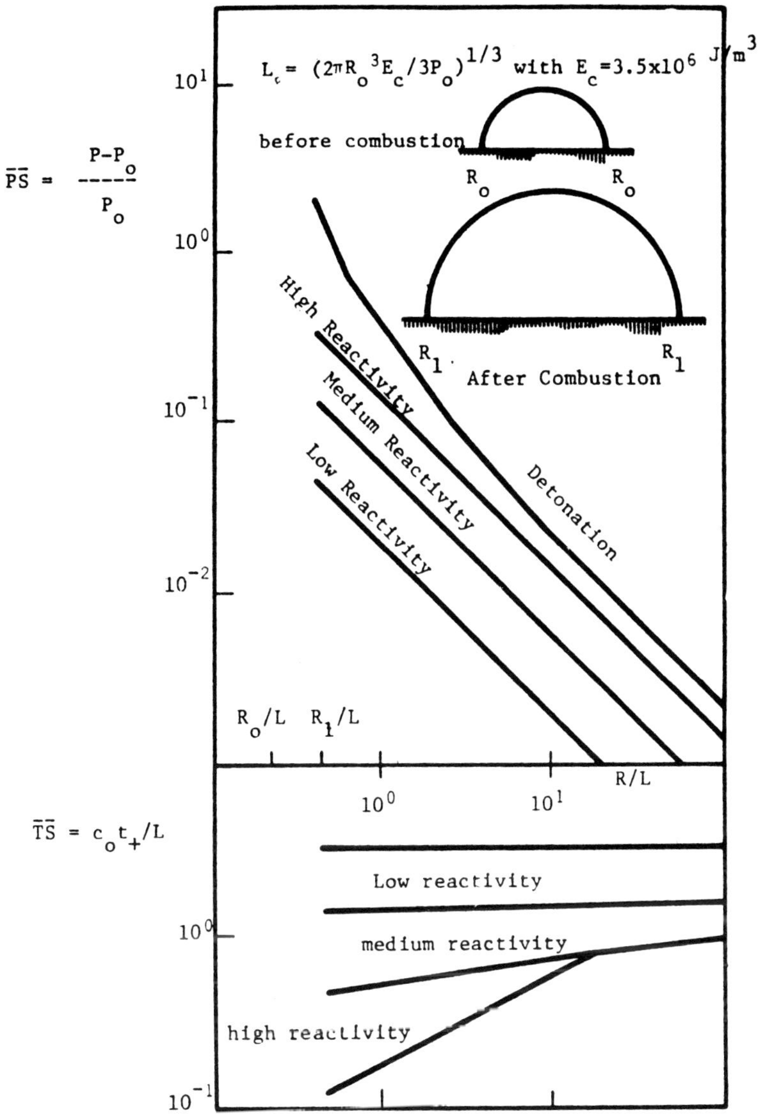

Fig. 3-13. TNO model of the blast effects produced by UVCEs [Wiekema (1980)].

The characteristic equations together with boundary conditions describe and determine the flow field once the energy release rate is given. The formation and location of the shock waves in that flow field is then solved by the adoption of the "area balancing technique" described by Witham. The results calculated by this characteristic method are illustrated by the relationship between the peak overpressure of the shock wave $\bar{P}_s = (P - P_o)/P_o$ and average flame velocity $\bar{V}_{f1}$.

Before the results can be given, the characteristic explosion length must be defined as:

$$L_o = \left(\frac{V_o E_c}{p_o}\right)^{1/3} = \left(\frac{2}{3}\pi R_o^3 \frac{E_c}{p_o}\right)^{1/3} \tag{35}$$

where E_c is the total heat release per unit volume and E_c is approximately 3.5×10^6 J/m^3 for common hydrocarbon fuels.

Now, the results are listed below:

average flame velocity $\bar{V}_{f1}$ (m/s)	peak overpressure $\bar{P}_s = (P-P_o)/P_o$
40	$2 \times 10^{-2}\ L_o/R$
80	$6 \times 10^{-2}\ L_o/R$
160	$15 \times 10^{-2}\ L_o/R$

The results are also plotted in Fig. 3-13 and the levels of reactivity correspond to average flame velocities: 40 m/s for low reactivity, 80 m/s for medium reactivity, and 160 m/s for high reactivity. The curve labelled "detonation" in Fig. 3-13 is fitted to the data of Kogarko, et al (1966) adjusted to the scaling and hemispherical geometry.

The positive phase duration of the blast wave should be the time interval of the arrival of the leading shock front and the arrival of the zero overpressure point behind the shock which propagates with the ambient sound speed.

In Fig. 3-13, the word "reactivity" means "the susceptibility to a flame acceleration", i.e., in identical situations, higher flame acceleration can occur and hence higher flame velocity can result in combustible mixtures of higher level of reactivity. For instance, in identical situations the blast damage of an acetylene-air explosion will be more serious than that of a methane-air explosion due to the greater flame acceleration and higher flame speed which can occur in the case of acetylene.

The fuels are divided into reactivity classes on the basis of experience and the reactivity classes of various hydrocarbon fuels are listed in Tables 3-14 and 3-15. In these tables those gases and vapors about which little or nothing is known are all classified in "high reactivity" and are indicated by "*".

TABLE 3-14. TNO Classification of Shock Wave Reactivity [Wiekema (1980)]

Explosion limits, flash point, and autoignition temperature of the various gases with a boiling point lower than 21°C together with reactivity.

Gas	Explosion Limits (vol. %)	Flash Point (°C)	Autoignition Temperature (°C)	Shock Wave Model Reactivity
Acetaldehyde	4-57	-50	140	average
Ammonia	15-28	< -33	630	low
1,3 Butadiene	1.1-12.5	-85	415	average
n-Butane	1.5-8.5	-138	360	average
Propane	2.1-9.5	-187	465	average
Propene	2-11.7	<- 48	455	average
1-Butene	1.6-10	-185	380	average
Dimethylamine	2.8-14.4	< -18	400	average
Ethyl Chloride	3.8-15.4	-58	510	low
Ethane	3.0-13.5	< -89	510	average
Ethene	2.7-34	< -104	425	average
Ethylene Oxide	3-100	-18	440	high
Methane	5-15	< -161	540	low
Methyl Bromide	8.6-20	< 4	530	low
Methyl Chloride	10.7-17.2	< -24	625	low
Vinyl Chloride	4-29	-77	470	average
Formaldehyde	7-73	-19	430	high*
Carbon Monoxide	12.5-74.2	< -191	605	low
Hydrogen Sulphide	4.3-46	< -60	270	high*
Acetylene	1.5-100	< -84	300	high[1]

[1] In view of the high reactivity of acetylene, it is advisable for this gas to use only the top line of the "high reactivity" range.

TABLE 3-15. TNO Classification of Shock Wave Reactivity [Wiekema (1980)]

Explosion limits, flash point, and autoignition temperature of the various liquids with a boiling point higher than 21°C together with reactivity.

Liquid	Explosion Limits (vol. %)	Flash Point (°C)	Autoignition Temperature (°C)	Shock Wave Model Reactivity
Acrylonitrile	3-17	-5	480	average
Acetonitrile	3-?	2	525	average
Allyl Chloride	3.2-11.2	-30	390	low
Carbon Disulphide	1-60	-30	100	high*
Diethylamine	1.7-10.1	< -20	310	average
Propylene Oxide	1.9-37	< -37	430	high*
Vinyl Acetate	2.6-13.4	-8	425	high*
Tetra Ethyl Lead	1.8-?	≈ 80	320	low
Allyl Alcohol	2.5-18	21	375	high*
Benzene	1-8	-11	550	high*
1,3 Dichloropropene	3.5-14.5	29	?	low
Epichlorhydrin	2.3-34.4	28	385	low
Ethylene Diamine	2.7-16.6	34	385	average
Ethyl Formate	2.7-13.5	-20	440	high*
Fromic Acid	14-33	69	> 520	average
Methyl Acrylate	2.8-25	-3	390	high*
Methyl Formate	5-23	< -20	450	high*
Solvent Naphtha	1.0-7.5	> 28	≈ 280	high*
Ethyl Mercaptan	2.8-18.2	< - 20	295	high*
t-Butyl Mercaptan	?	?	?	high*
n-Butyl Mercaptan	?	?	?	high*
Tetrahydrothiophene	?	12.7	?	high*

The selection of the value of parameters, or the location in the band depends upon the situations in the surroundings. If there is a probability of a high degree of flame acceleration, due to the presence of many obstacles or partial confinements in the cloud, then the upper value in the band must be taken. If, on the other hand, there is no reason to expect a substantial flame acceleration, the lower value in the band is taken. If one is not certain about the occurrence of flame acceleration, then either the upper value is used as the worst case evaluation, or the middle point between the upper and lower values can be taken as the average.

3) The multi-energy method

The multi-energy method proposed by van den Berg, et al (1989, 1985) is based on the concept that strong blast is generated only in places characterized by a considerable degree of partial confinement while other, usually large parts of the cloud burn out without any significant contribution. Therefore the blast produced by a vapor cloud explosion will consist of several blast waves propagating independently with each one of them being modeled by an equivalent hemispherical combustible mixture. The basic tool of the multi-energy method is the relations of the energy-scaled blast wave properties as the function of the scaled distances as shown in Fig. 3-14, where solid lines represent high strength blast while dashed lines represent low strength pressure waves which may steepen to shock waves in the far field. The initial strength of the blast is a variable and has been indicated with a number ranging from 1 for insignificant blast up to 10 for a gas detonation. Another basic feature of the blast-distance relations is that at a certain distance the blast is nearly independent of its initial strength when the initial strength is 6 or higher.

According to Van Wingerden, et al (1989), the multi-energy method is employed as follows: The blast from each congested area inside a vapor cloud in a chosen accident scenario is represented as the blast from a hemispherical fuel-air cloud the volume of which is equivalent to the volume of the congested area. Blast parameters in the vicinity of the cloud can be estimated using Fig. 3-14 providing that an initial blast strength must be chosen and the amount of combustion energy involved must be substituted.

A demonstration of the method is given below using the Flixborough incident in 1974 since it has been best documented. According to the multi-energy concept the blast of the Flixborough explosion can be modeled by the blast of two fuel-air charges: a low-strength charge corresponding with the unconfined parts of the cloud and a high-strength part corresponding with the congested parts in the cloud. The initial strength of the partially confined parts of the cloud has been assumed to be 7 while that from the unconfined parts of the cloud is assumed

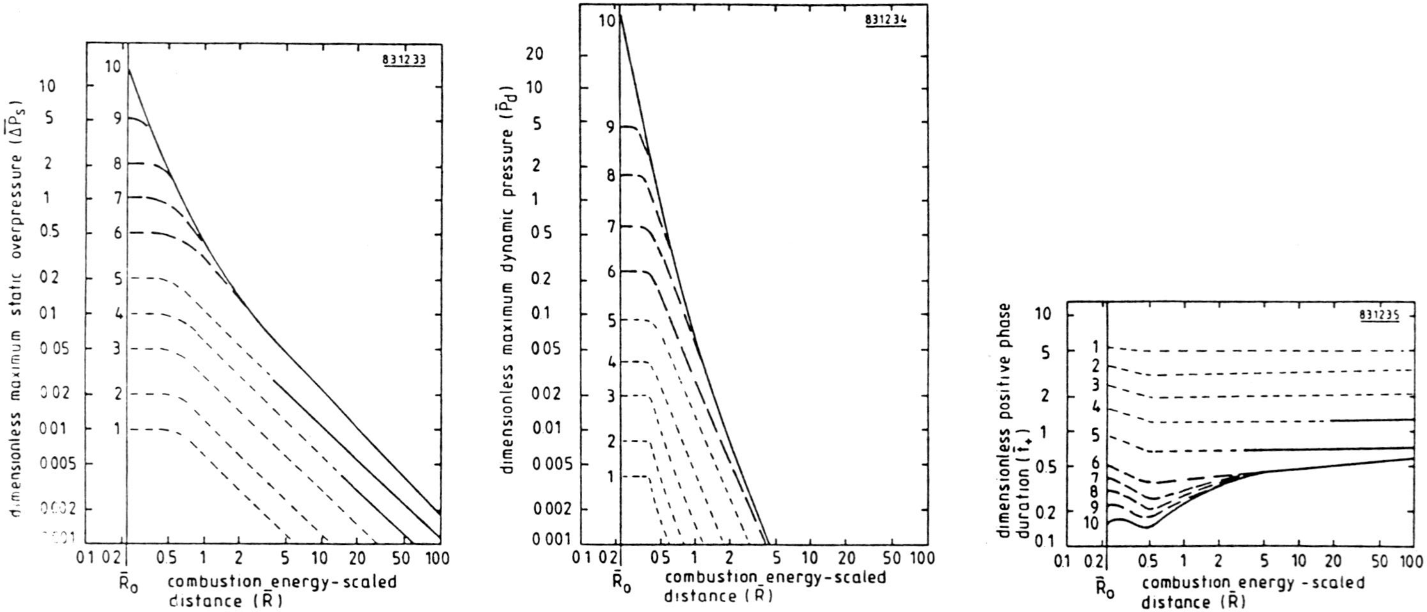

Fig. 3-14. Blast-distance reference relations for stoichiometric hydrocarbon-air mixtures [Van Wingerden, et al (1989)].

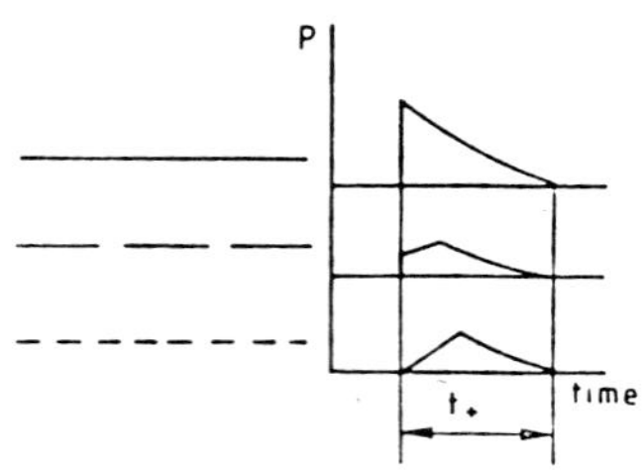

$$\overline{\Delta P_s} = \frac{\Delta P_s}{P_o} \; ; \; \bar{P}_d = \frac{P_d}{P_o} \; ; \; \bar{t}_+ = \frac{t_+ c_o}{(E/P_o)^{1/3}} \; ; \; \bar{R} = \frac{R}{(E/P_o)^{1/3}}$$

P_o = atmospheric pressure

c_o = ambient velocity of sound

E = combustion energy participating in the explosion

R_o = fuel-air charge radius

to be 2. The resulted blast parameters are illustrated in Fig. 3-15 along with the data derived from a blast damage analysis according to Sadee, et al (1977).

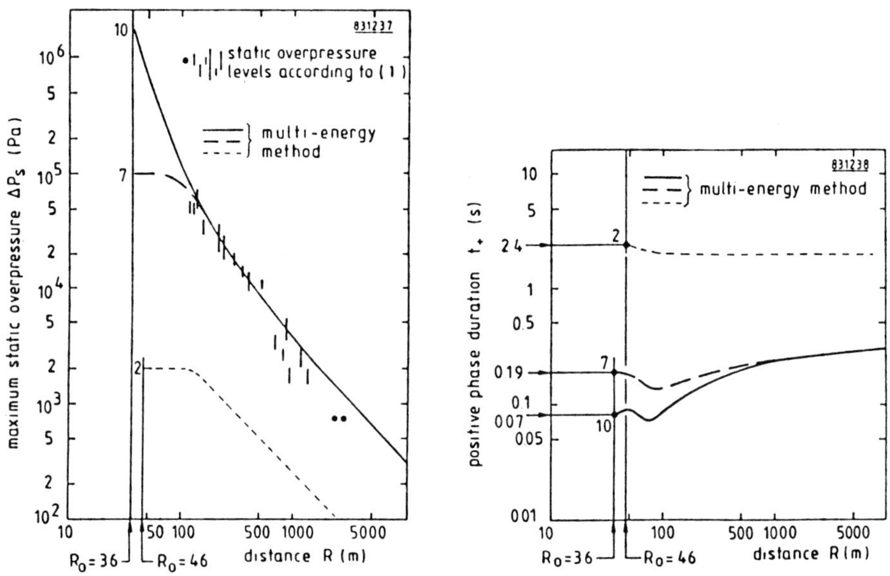

a) Maximum static overpressure b) Positive phase duration

Fig. 3-15. Representation of the blast from the Flixborough explosion [Van Wingerden, et al (1989)].

A consistent picture is obtained by assuming that only about 30% of the total amount of cyclo-hexane release was involved in the strong explosive combustion. The positive phase duration of the shock wave generated by the strong explosion increases up to 0.2 seconds in the far field which is in good agreement with the estimation by Roberts and Pritchard (1982). It is obvious that the combustion of the large unconfined parts of the cloud which took several seconds hardly contributed to the blast effects.

It is clear that the initial strength of a blast wave is a function of gas reactivity, obstacle parameters and the degree of confinement. There has been, unfortunately, no quantitative prediction of the initial strength available so far and the choice of the initial strength number could only be in qualitative terms which will be discussed briefly as follows:

Experimental investigations showed that for relatively low flame speeds the laminar burning velocity can be used to estimate the effects of different gases and vapors in the same configuration of obstacles. It has been found that the course of flame propagation in the same obstacle configuration for all gases falls into a single curve, providing that the flame speed is nondimensionalized using the respective laminar burning velocity as shown in Fig. 3-16 [Van Wingerden and Zeeuwen (1983)].

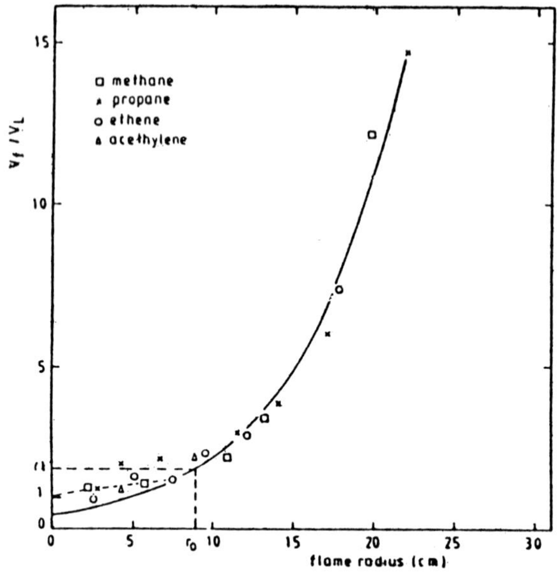

Fig. 3-16. Flame speeds dependent on the flame radius in a vertical obstacle environment for various mixture reactivities. The flame speeds have been non-dimensionalized using the respective laminar flame speeds [Van Wingerden and Zeeuwen (1983)].

The two most important obstacle parameters that can be distinguished are blockage ratio (ratio of area blocked by obstacles and total area) and the pitch (the relative distance between two successive obstacles or obstacle rows). It has been found that the flame speeds and hence the pressure development increase with the increase of the blockage ratio [Moen, et al (1981, 1982); Van Wingerden, et al (1983); Hjertager, et al (1984)]. With regard to the influence of obstacle pitch, in tubes the optimum pitch value for baffle plates seems to be between 5.4 and 7.5 times the baffle plate height [Moen, et al (1982); Hjertager (1984)]. For double plate configurations the optimum seems to exceed 6 obstacle diameters [Van Wingerden (1989)].

Experiments performed by Van Wingerden and Zeeuwen (1983) in a double plate large scale obstacle array showed the dramatic influence of confinement. In case of a confining plate over the obstacles, flame speeds up to 420 m/s were observed for ethylene, whereas with no top plate the maximum flame speed was only 24 m/s. Similar results were found for various other gases [Van Wingerden (1983)].

In summary of the deflagration models, the prediction of blast effects using Strehlow's spherical flame model requires a specified flame speed which itself remains an unanswered question. The TNO hemispherical flame model, on the other hand, does not require the specification of flame speed but has based its prediction on the classification of combustible gas. The classification according to the reactivity of the gas which is not a well defined property and the prediction of blast effects using the TNO hemispherical model has not taken the effect of confinements into account, which has been proven extremely important. Finally, in the multi-energy method, partial confinement has been recognized as a factor of primary influence on both the scale and strength of gas explosions. However, the procedure of the prediction of blast effects from UVCEs using the multi-energy method is still far from complete. For instance, the concept of partial confinement should be defined and parameterized, and the concept of separate modeling of the blast generating objects should be verified [Van den Berg, et al (1989)].

4. Blast effects produced by vapor cloud detonations

Due to the huge amount of energy required for the direct initiation of detonation in most fuel-air mixtures, this mode of detonation initiation is unlikely to occur accidentally. However, the estimation of the blast effects produced by a detonating cloud is still of practical importance for two major reasons. First, it is possible for a detonation to be realized via the mechanism of flame acceleration and the transition from deflagration to detonation, or the initiation by a hot turbulent jet resulted from explosion venting. Second, the assumption of the establishment of a detonation always gives the estimation of the worst-case blast wave properties which may sometimes be required for the conservative design of blast-resistant structures or choice of safety distances.

The prediction of the blast effects generated by vapor cloud detonations has already been illustrated in the above deflagration models as the extreme cases; for instance, flame Mach number equals C-J detonation Mach number in Strehlow's spherical model (see Fig. 3-11), the detonation curve of reactivity in Fig. 3-13 for TNO hemispherical model and the highest initial strength, i.e., number 10, in Fig. 3-14.

There have been a number of blast measurements made from detonating premixed gas mixtures in spherical fuel-air or hemispherical on the ground configurations. Brossard, et al (1984) collected previous data and conducted many more such tests, with gas volumes ranging from about 5×10^{-4} m^3 up to 1.45×10^4 m^3. These authors scaled the data for a number of blast parameters according to a reduced form of the scaling law given by Equation (17) (i.e., ignoring parameters describing scaled rate of energy release and source shape). They developed probably the first available set of scaled curves based on experiment for such detonations. Their scaled data, including the important negative-phase properties, are shown here as Figs. 3-17 through 3-22. Smooth curve fits to the data for positive-phase blast wave properties and positive plus negative phases appear in Figs. 3-23 and 3-24.

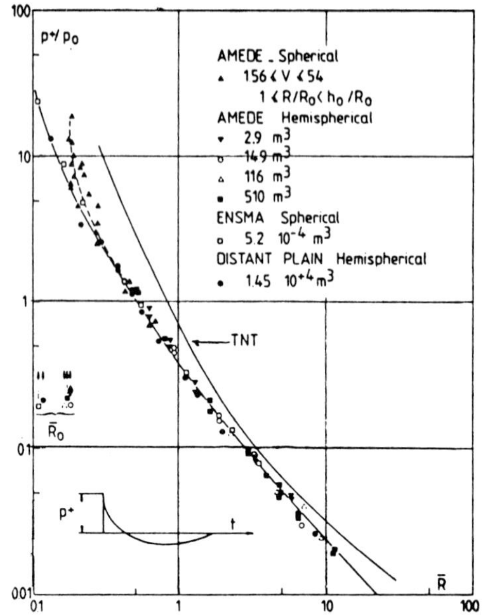

Fig. 3-17. Reduced peak overpressure vs reduced distance [Brossard, et al (1984)].

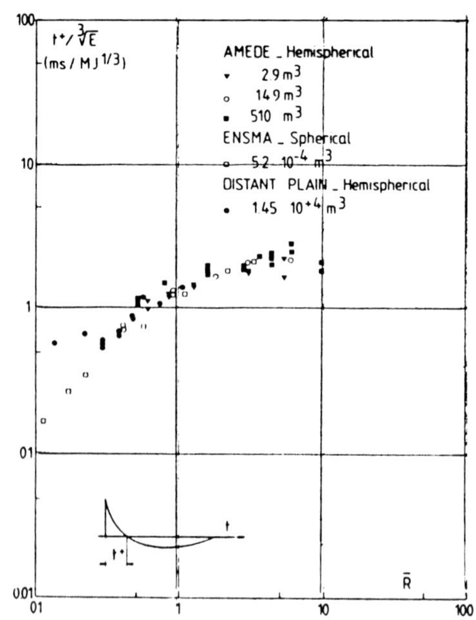

Fig. 3-18. Reduced positive duration vs reduced distance [Brossard, et al (1984)].

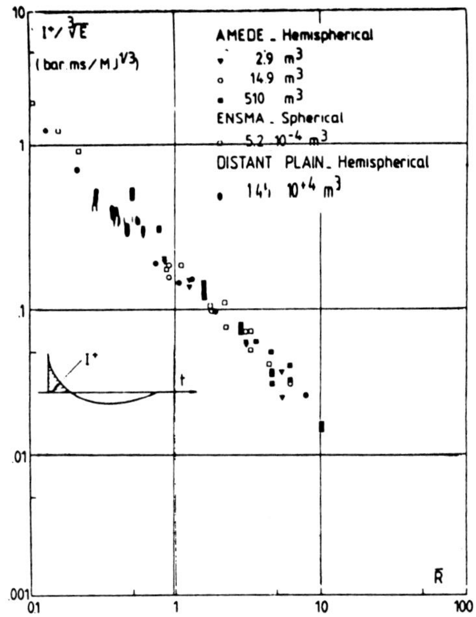

Fig. 3-19. Reduced positive specific impulse vs reduced distance [Brossard, et al (1984)].

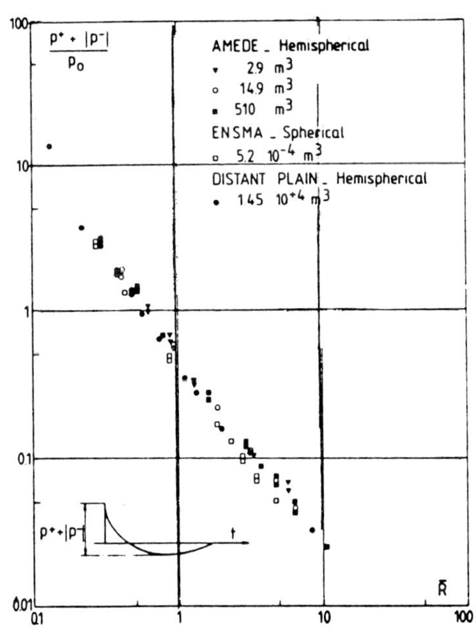

Fig. 3-20. Reduced total amplitude of the pressure signal vs reduced distance [Brossard, et al (1984)].

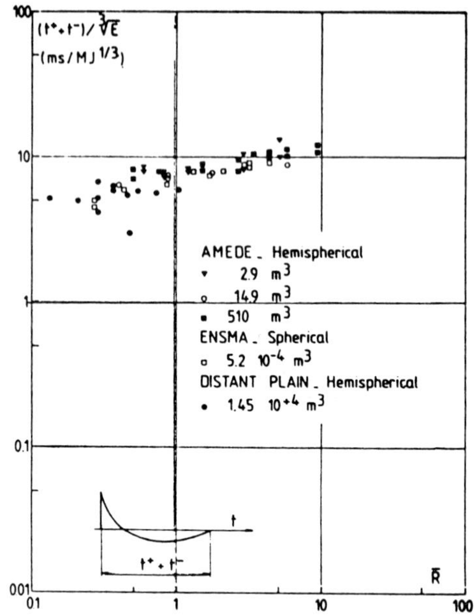

Fig. 3-21. Reduced total duration of the pressure signal vs reduced distance [Brossard, et al (1984)].

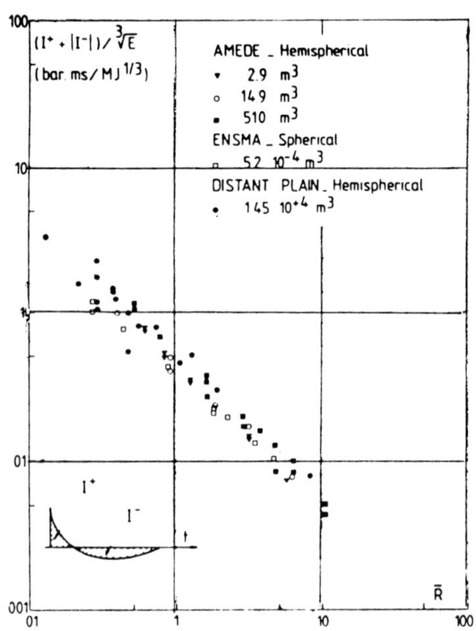

Fig. 3-22. Reduced total absolute specific impulse vs reduced distance [Brossard, et al (1984)].

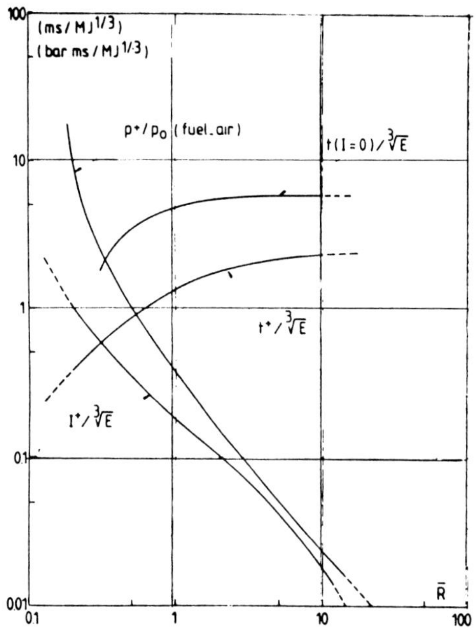

Fig. 3-23. Characteristics of the positive phase of the pressure profile vs reduced distance [Brossard, et al (1984)].

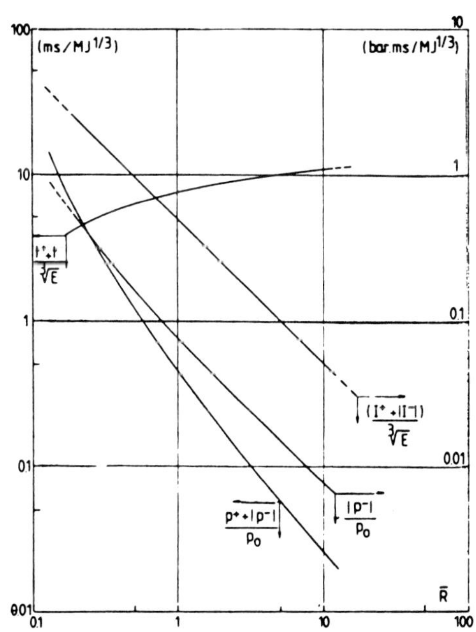

Fig. 3-24. Characteristics of the pressure profile, taking into account the negative part, vs reduced distance [Brossard, et al (1984)].

Although these authors note the occurrence of the second shock, they do not show it in their idealized pressure profiles, nor do they report measurements of second shock properties. Also, please note in Figs. 3-17 through 3-24 that a number of blast parameters are not rendered nondimensional, even though they are plotted versus nondimensional distance. Throughout, effects of a hemispherical charge on the ground with radius R_o were assumed to be the same as for a sphere of radius R_o in fuel air. So, source energy was always

$$E=\frac{4\pi R_o^3}{3}\rho_f E_f \tag{36}$$

where ρ_f and E_f are density and heat of combustion for the specific gas mixture. This gives initial dimensionless cloud radius as

$$\overline{R}_o=\left[\frac{4\pi\rho_f E_f}{3p_o}\right]^{-1/3} \tag{37}$$

For hydrocarbon-air mixtures, $\overline{R}_o \approx 0.18$, while for hydrocarbon-oxygen mixtures, $\overline{R}_o \approx 0.12$. These authors give least-squares fits to positive phase blast wave properties:

$$\ln(p^+/p_o)=-0.9126-1.5058(\ln\overline{R})+0.1675(\ln\overline{R})^2-0.0320(\ln\overline{R})^3 \tag{38}$$

$$\ln(t^+/\sqrt[3]{E})=+0.2500+0.5038(\ln\overline{R})-0.1118(\ln\overline{R})^2 \tag{39}$$

$$\ln(I^+/\sqrt[3]{E})=-1.5666-0.8978(\ln\overline{R})-0.0096(\ln\overline{R})^2-0.0323(\ln\overline{R})^3$$

$$\text{for}\quad 0.3\le\overline{R}\le 12 \tag{40}$$

Note that Equation (36) is nondimensional, while Equations (37) and (38) are dimensional, with units shown in Fig. 3-19.

The one-dimensional methods, however, have not included the influence of cloud geometry which is usually irregular in shape for actual vapor cloud explosions. So there have been some two-dimensional blast analyses designed to simulate blast effects of real cloud geometries. The explosion cloud can perhaps be approximated by elongated cigar-shaped or flat disc-shaped clouds. The approximation of flat disc-shaped or pancake-shaped clouds on the ground surface has often been used because vapor densities of most fuels are greater than air.

Also, the contact of the cloud with the ground maximizes the blast damage for a given amount of energy release due to the reflection of the blast wave from the ground.

Sichel (1979) presented an analytical model of planar detonation with side relief to calculate the pressure history on the ground during the detonation of a cloud with small height-to-diameter ratio. Fishburn, et al (1981) has carried out a numerical study using the two-dimensional HEMP computer code to simulate centrally initiated detonation in a thin disk shaped cloud of 4.57m thick and 128m in diameter and consisting of stoichiometric kerosene-air mixture. Pressure and impulse histories within a detonation cloud of stoichiometric heptane aerosol of 70m in diameter and about 5.8m thick have been measured experimentally [Fishburn, et al (1981)]. The height-to-diameter ratio and the fuel of the heptane cloud are not the same as that assumed in the numerical study; however, the ratio was sufficiently small and the detonation parameters of stoichiometric kerosene-air and heptane-air are close enough that similar results could be expected.

The pressure histories on the ground at the distance within the cloud obtained by the above analytical model, numerical calculation and experimental measurement are compared in Fig. 3-25. The peak overpressure obtained from the numerical calculation is obviously lower than the theoretical and experimental results. This is due to the use of relatively large computational cell size which smeared out the combustion region. Nevertheless, the calculated impulse is not greatly affected as shown in Fig. 3-26 and the general agreement between the calculation and experiments is quite good.

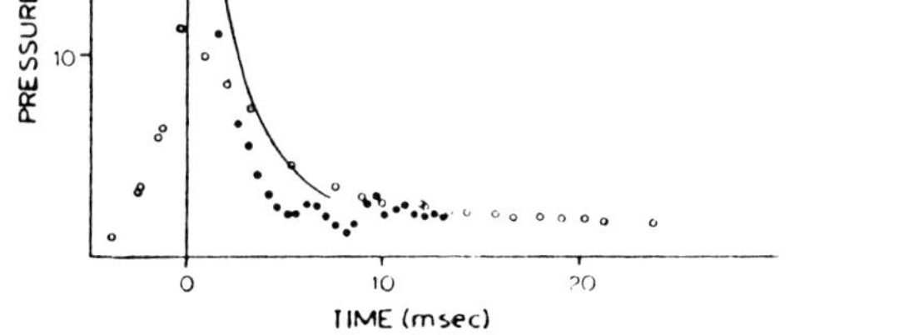

Fig. 3-25. Pressure history within pancake shaped cloud [Sichel (1979)].

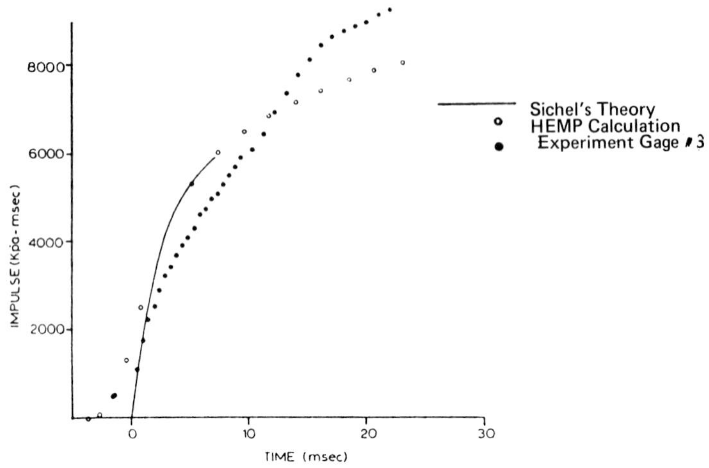

Fig. 3-26. Impulse history within pancake shaped cloud [Sichel (1979)].

According to Baker (1984), the most systematic numerical study to date of blast from detonating pancake shaped clouds is that of Raju and Strehlow (1984). The computations were made using an axisymmetric hydrodynamic code for detonations in clouds in free air with aspect ratios of 5 and 10. The results of the dimensionless peak overpressure and first positive phase specific impulse along three radial lines from the cloud center are reproduced here as Figs. 3-27 to 3-30. For comparison, scaled curves for free-air detonations of spherical Pentolite are also plotted in these figures.

Peak side-on overpressure, scaled specific impulse, scaled time of shock arrival and scaled positive phase duration are plotted as functions of scaled distance by Baker (1984) according to Hopkinson-Cranz scaling as shown in Fig. 3-31. The empirical curves are based on experimental blast data reported by Axelsson, et al (1978) and Kiwan, et al (1975) for detonation of pancake-shaped ethylene oxide fuel-air clouds with an aspect ratio of about 5, which are FAE clouds formed by explosive dispersion from cylindrical containers. The distance is measured along the midplane through the pancake cloud, and so corresponds to the maximum of blast properties.

A two-dimensional axisymmetric numerical study for the blast effects from the detonation of a pancake-shaped cloud has also been carried out by Wiedermann and Eichler (1981). The results, which are labeled "LNG cloud", show that the peak overpressure within the cloud is the detonation overpressure of 1.45 MPa for a stoichiometric natural gas-air mixture and outside the edge of the cloud the peak overpressure decreases rapidly. A value of 50 kPa is reached at approximately 15 height units and an extrapolation indicates that the 7 kPa (1 psi) level, which is of interest in structural response and personnel safety, occurs at a range of approximately 3.6 km (about 100 height units). The calculation has been made for a cloud of 1.1 km in radius and 25m in height, but the results are presented in terms of the cloud height unit, hence possess some universality.

In conclusion, the large blast effects from fuel-air explosives may be attributed to the fact that heats of combustion of fuels are generally much larger than heats of explosion of condensed explosives, e.g., the heat of combustion for kerosene, 44.3 MJ/kg, is about 10 times the heat of explosion for TNT. In addition, the tendency of the clouds to form a thin disk and creep along the ground can significantly enhance the blast effects, e.g., 20 times the fuel weight in TNT. Hazard assessments should include this factor when the estimation of the worst case is required.

5. Experimental investigations on flame acceleration

The results of a great number of experimental studies, both in laboratory and large scale, have shown that flame acceleration processes cannot become severe if the vapor cloud is completely unconfined. In a series of experiments conducted by Lind (1975) and Lind and Whitson (1977) in 5m and 10m radius hemisphere plastic bags mounted on a smooth concrete pad, no overpressures above 0.1 bar, which is the lowest pressure that could be recorded by the instrumentation system, were recorded. The fuels tested were methane, ethylene, acetylene, propane, butadiene and ethylene oxide at near stoichiometric concentrations. The amount of fuel tested for each trial ranged from 21 kg (5m radius hemisphere) to 290 kg (10m radius hemisphere). Although an increase of normal burning velocity to a factor of 2.1 for methane and 1.7 for the other fuels were observed (the factor of 0.8 for butadiene is obviously spurious), in no case did the flame continue to accelerate to a higher velocity. More discussion of these tests appears in Baker, et al (1983). Similar results were obtained for large scale LNG spill test at China Lake in California, in 1981 [Rodean, et al (1984)]. This is reasonable since the blast effects are caused by the massive increase in the volume of gas. If this volume production is released without restraint at a speed much less than the sound velocity in the surrounding atmosphere, the blast effect is negligible and the heat radiation will be the only important form of energy release which determines the damage.

There are, however, obvious discrepancies between the lack of blast wave development for the relatively large scale balloon tests and accident investigation which indicates evidently strong and devastating blast waves. These discrepancies are probably attributed to the very clean and unobstructed configuration for the plastic bag tests and the very complex geometry existing in most of the plants experiencing damaging UVCEs. It is thus clear that some types of flame acceleration mechanisms are critical to the development of UVCEs.

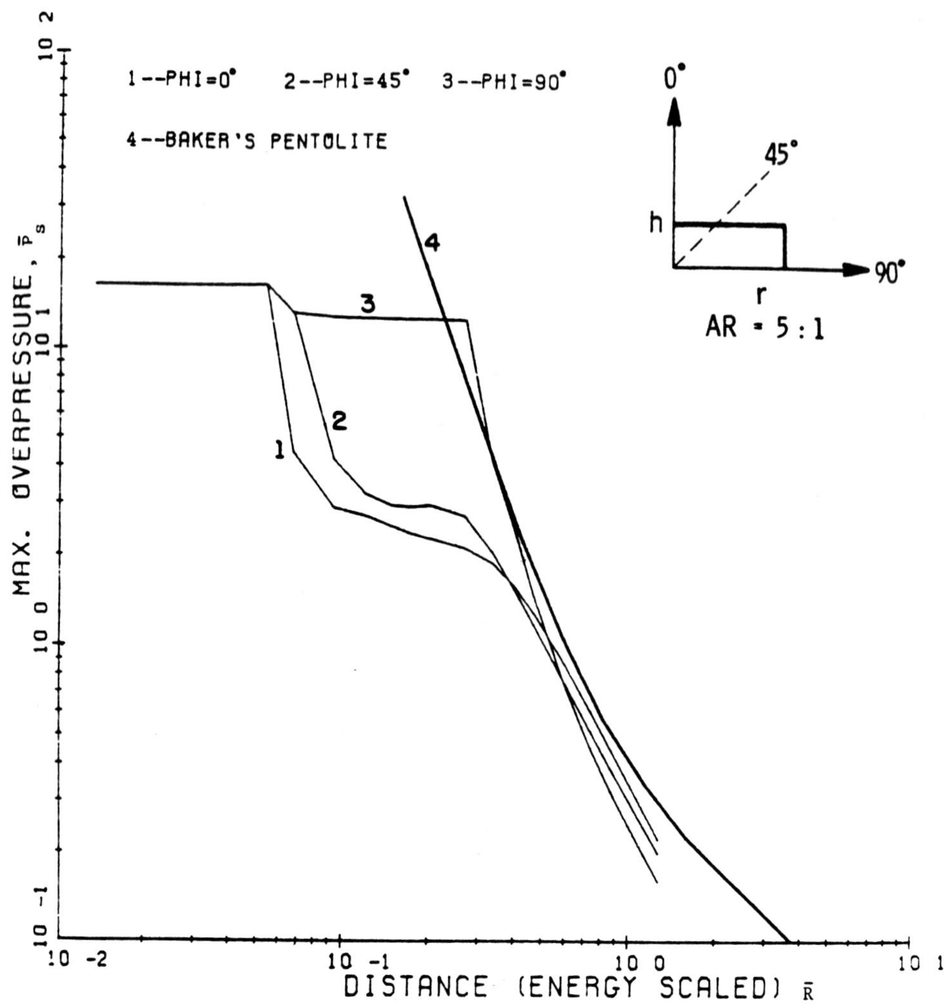

Fig. 3-27. Dimensionless peak overpressure versus energy-scaled distance, detonation of pancake cloud with aspect ratio of 5:1 [Raju and Strehlow (1984)].

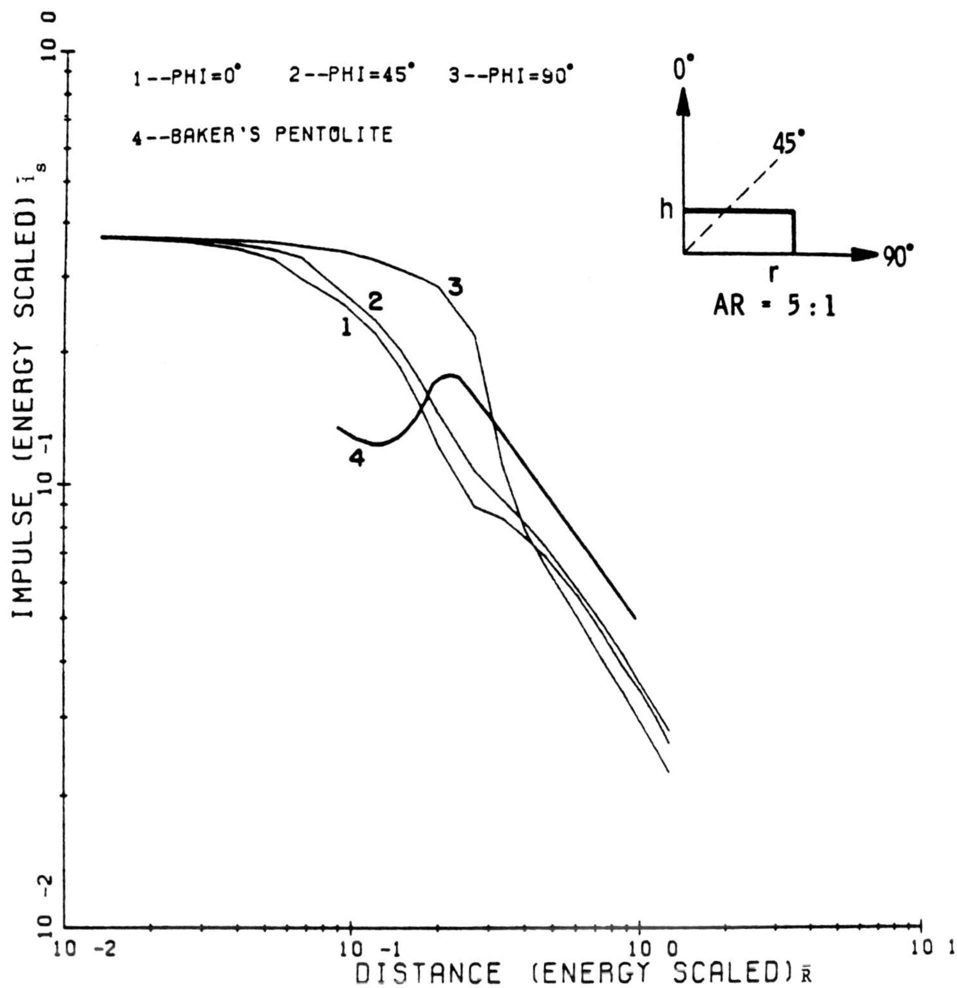

Fig. 3-28. Dimensionless specific impulse versus energy-scaled distance, detonation of pancake cloud with aspect ratio of 5:1 [Raju and Strehlow (1984)].

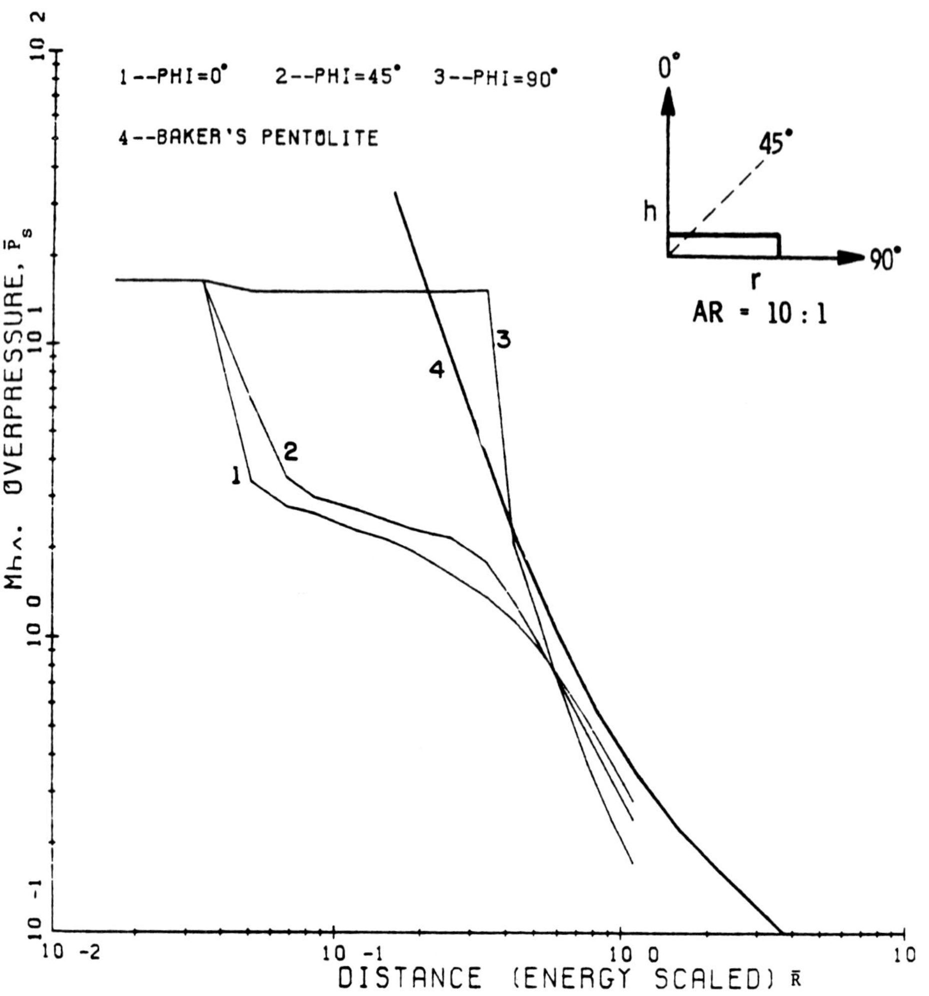

Fig. 3-29. Dimensionless peak overpressure versus energy-scaled distance, detonation of pancake cloud with aspect ratio of 10:1 [Raju and Strehlow (1984)].

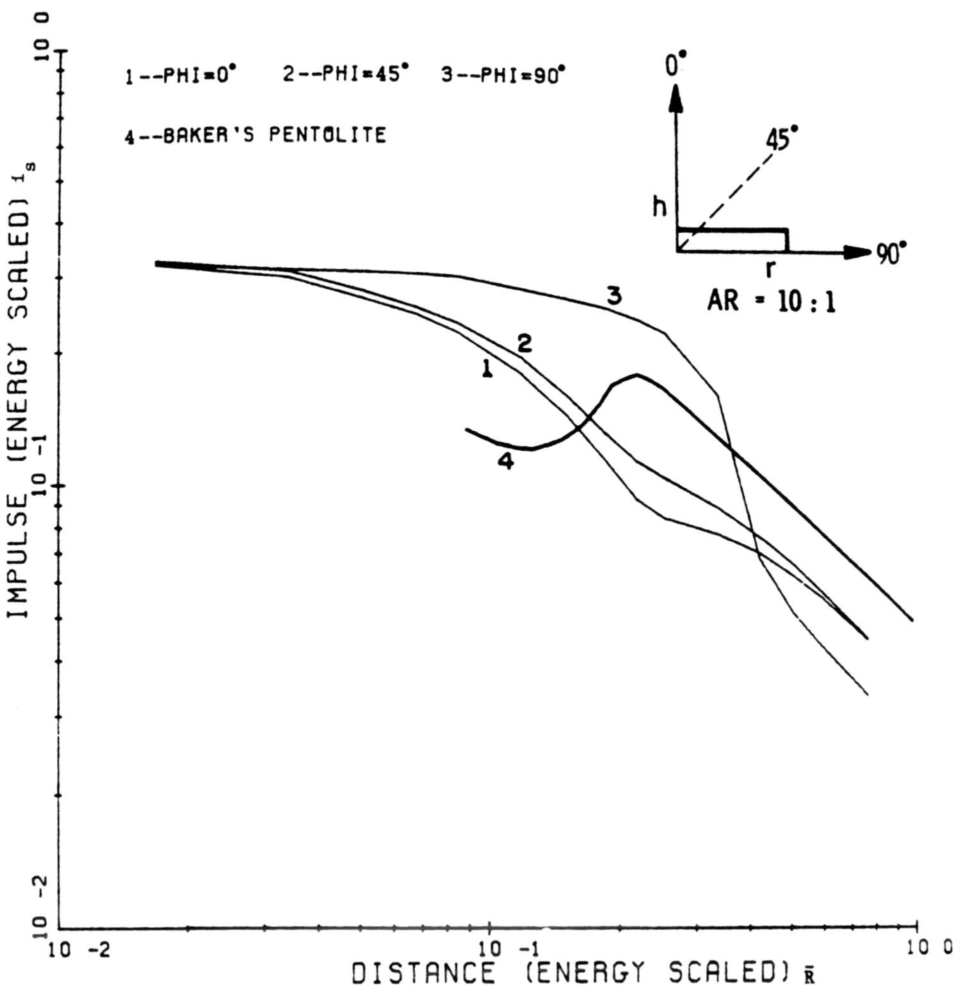

Fig. 3-30. Dimensionless specific impulse versus energy-scaled distance, detonation of pancake cloud with aspect ratio of 10:1 [Raju and Strehlow (1984)].

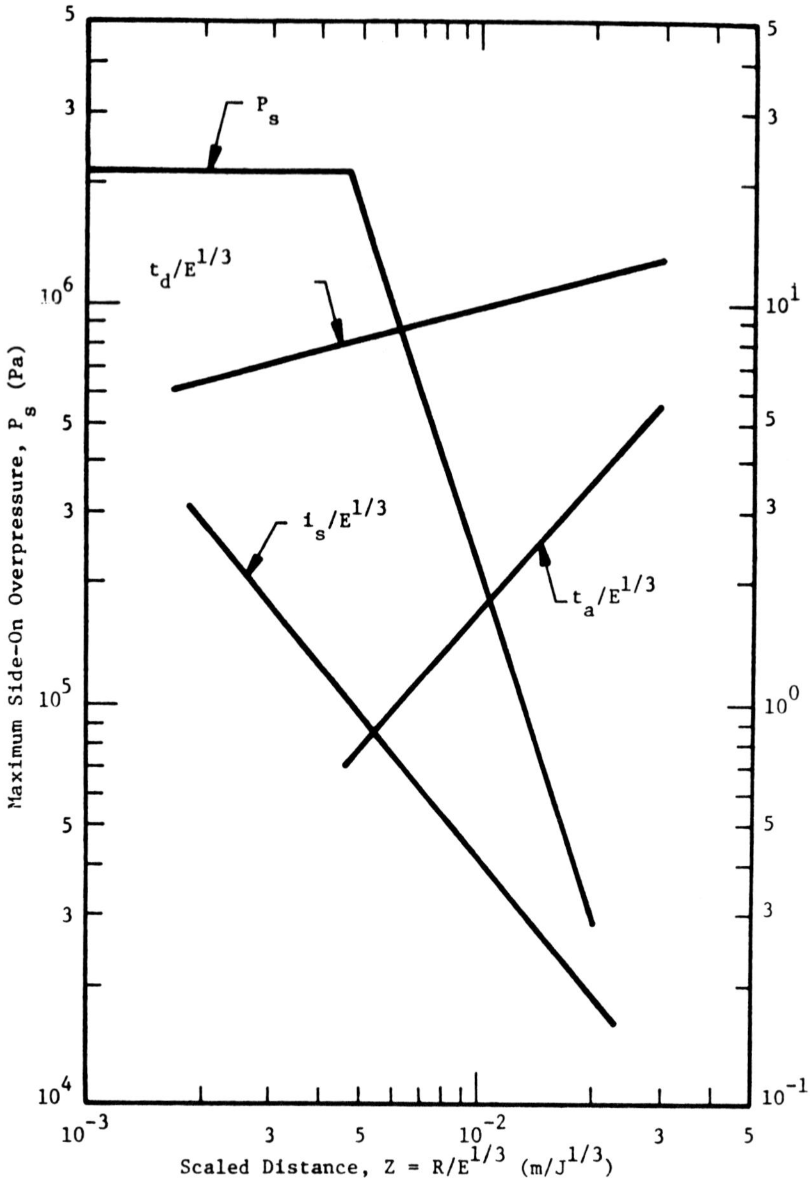

Fig. 3-31. Scaled FAE blast curves [Baker (1984)].

Several mechanisms of flame acceleration were proposed, such as self-turbulence of the flame (flame acceleration due to flame instabilities), the Markstein-Taylor instability (flame acceleration due to flame-shock interaction) and flame propagation by heat radiation (dust particles suspended in the flow ahead of the flame are heated by radiation from the flame and then ignite the reactive mixture there, resulting in a large effective flame front area), etc. Among the various mechanisms, the most effective appears to be the flame acceleration due to interaction with obstacles, which is especially effective in partially confined areas. Both laboratory tests and large scale field tests have revealed that flame behavior is strongly affected by the presence of obstacles and partial confinements.

Recent field experiments on flame propagation through very large obstacle arrays have confirmed the important effect of obstacles [Harrison and Eyre (1986)]. These tests, organized by Shell Research Ltd., Thornton Research Center, are the largest performed so far on vapor cloud explosions, involving 4000m^3 of premixed natural gas-air and propane-air mixtures. A sketch of the experimental layout is shown in Fig. 3-32. The pressure records showed that each grid caused acceleration of the flame, but the peak pressure developed within the rig was relatively modest (less than 0.7 bar). The results confirmed that deep obstacle arrays and/or considerable confinement are required if significant blast effect is to result from the weak ignition of a flammable cloud.

Large scale experiments performed in Norway [Moen, et al (1982)] showed that with a worst-case geometry, e.g., a pipe containing ring obstacles, and the flame propagating away from the closed end towards an open end, extremely violent explosions, perhaps a detonation, can occur even in a methane-air mixture. Nevertheless, it must be stressed that as soon as a fast flame emerges from a confined or congested region into an open area, the overall flame speed drops rapidly.

In order to assess the potential for flame acceleration in industrial environments, where process equipment such as vessels and pipelines are present, a series of tests were performed in a top-vented channel, 1.8m x 1.8m in cross-section and 15.5m long with repeated obstacles by Moen, et al (1986). Tests were performed with acetylene, propane and hydrogen sulphide fuels mixed with air and repeated cylindrical obstacles of two diameters (500 and 220 mm) mounted across the channel at regular intervals (spacing equals 1.27m and 0.63m). In near stoichiometric acetylene-air mixture, the flame accelerates as it propagates down the channel and reaches speeds up to 400 m/s prior to the occurrence of localized explosions which trigger the onset of detonation. The behavior of flames in lean acetylene, propane and hydrogen sulphide-air mixtures is much less dramatic with flame speeds ranging from 25 m/s up to 200 m/s and associated pressures less than 50 mbar. However, the continuous flame acceleration seen in more confined

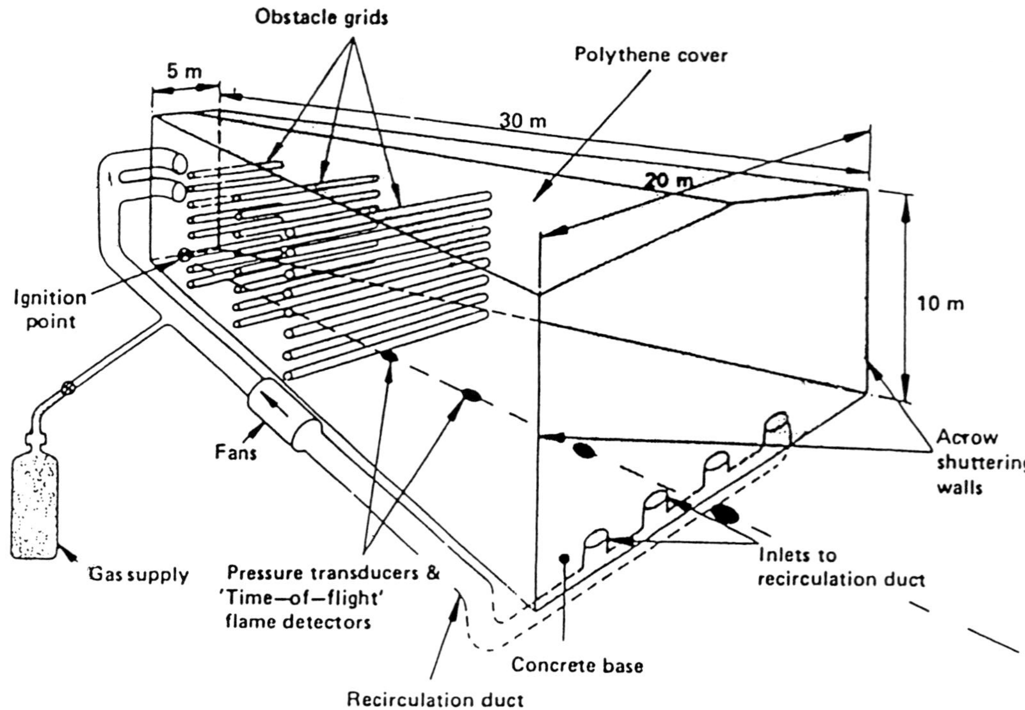

Fig. 3-32. Schematic diagram of experimental rig.

configurations is not observed in the present configuration with these fuel-air mixtures.

In view of the importance of the obstacle induced turbulence on flame acceleration a great number of investigations have been made on this subject in the past decade [Moen, et al (1981), Wagner (1982)]. However, most of these investigations were experimental studies performed on a laboratory scale, aimed at studying the influence of repeated obstacles and vertical confinement. On a large scale, field tests were conducted [Zeeuwen, et al (1983)] with a cloud up to 100 kg of propane and the obstacles were placed in the area to be covered by the flammable cloud.

In these tests, liquid propane, up to 40 kg/sec, was evaporated from a pond and dispersed freely into the air. Depending on the weather conditions a certain part of the vapor will be mixed in a flammable proportion with air which was calculated by means of a dispersion model. It was found from the video recordings that the area bounded by the LFL (lower flammability limit) contour (the maximum distance covered by the flame) was more than 130m long and more than 75m wide for clouds up to 1000kg of propane. The height of the flammable part of the cloud was about 2m. Near the pond a large part of the mixture was too rich to be flammable. For tests without obstacles the flame speed was constant over large distances. Usually the flame speed was 3 to 10 m/s which is higher than that of 3-5 m/s measured on a small scale [Van Wingerden, et al (1983)]. Only in one trial, the high flame speed of 32 m/s (in one direction) was recorded. Tests with horizontal obstacles, six pipes of 0.3m diameter and one pipe of 0.6m diameter placed perpendicular to the wind direction, showed that in front of the obstacles the flame accelerated to a maximum speed of 16 m/s, but after passing the obstacles the flame speed dropped to a constant value of 6 and 10 m/s to the left and right respectively. In the tests with vertical obstacles, sewer pipes (average diameter 1m, length 2m) with a 3m pitch and maximum number of 104, the flame propagation was not influenced by the presence of the obstacles. In tests with vertical obstacles and vertical confinements with the top of the obstacle array covered by steel plates, the maximum flame speed amounted to 50-66 m/s. Within the array, the maximum pressure measured was 2.0 kPa (20 mbar). At 75m, it was 0.59 kPa (15.9 mbar) and the decay of the peak pressure with distance was stronger than expected from acoustic theory ($1/\bar{R}$ dependence).

It thus can be concluded that vertical confinement plays an important role in vapor cloud explosions since on an open site with vertical obstacles no flame acceleration occurs, while with horizontal obstacles flame accelerates only very slightly. Only within a vertically confined configuration the flame accelerates continuously and the blast effects measured are in agreement with the flame speeds.

IV. HYBRID EXPLOSIONS

A. Fundamentals

A number of very complex physical processes occur when combustible dusts and gases are mixed with air in strong, vented or unvented enclosures, and then ignited. The processes are dynamic (time-varying), and can include laminar and turbulent fluid flow, heat transfer from an ignition source to the dusts suspended in air and to the gas-air mixtures, combustion, flame acceleration by mechanisms which are poorly understood even for gas-air mixtures alone, and overall pressure and temperature increase and eventual decrease within the enclosure. The last two parameters determine whether the enclosure can survive the internal explosion.

Related work in the literature is, of course, voluminous. Here, we will highlight only those references which appear most pertinent.

Bartknecht (1978a) is an excellent general reference on pressures developed in vented and unvented enclosures by either internal gas or dust explosions. An empirically developed rule given in that reference notes that the maximum rate of pressure rise for explosion of a given combustible dust-air mixture or a given combustible gas-air mixture in a closed vessel, is inversely proportional to the cube root of the vessel volume. But, Bartknecht has very little discussion of confined explosions for what he calls "hybrid mixtures" of combustible dusts and gases.

Nagy and Mitchell (1963) report some data for coal-dust and gas explosions in an experimental coal mine. The work in these two references does show that the hybrid mixtures are more easily ignited than either one alone, and that the explosions are more violent.

Lee and Guirao (1982) is a very useful recent reference, including a number of papers on confined gas and dust explosions. Again, only one or two papers in the proceedings discuss hybrid explosions. Several do discuss in detail the role of turbulence in confined gas explosions, and identify parameters used to describe turbulent conditions.

Turbulent and buoyant effects in confined gas explosions have been treated in several other references, in addition to Lee and Guirao (1982). Nagy, et al (1971), discuss buoyant effects and the effects of vessel size for closed vessels while Yao (1974) discusses turbulent effects for vented enclosures. Finally, Makepeace (1978) has prepared an excellent review paper on the effect of turbulence on gas combustion.

B. Scaling

In performing a similitude analysis related to the physics of confined explosions of hybrid (gas-dust) combustible mixtures in air, we were largely guided by a previous analysis by Baker, et al (1980). That analysis was limited to dust explosions, but did include a number of parameters related to the dust, to the ignition source, to the enclosure, and to initial conditions, as well as response or output parameters. Combustion dynamics, available energies and heat transfer parameters were included, as were parameters describing enclosure venting. But, no turbulence parameters were included.

In the present analysis, the venting parameters are omitted, and a number of additional quantities for characterizing gas explosions are added to the parameters in Baker, et al (1980), including several for turbulent effects. The list of physical parameters we have assumed to be adequate to bound the problem is given in Table 4-1, with a description of each parameter and its dimensions in a Mass, Length, Time, Temperature (M, L, T, θ) system. There are 40 parameters.

Following methods outlined in Baker, et al (1973), the 40 physical parameters can be cast in the format 40-4=36 dimensionless ratios, or Pi terms, as shown in Table 4-2. To do this, we chose to nondimensionalize in terms of the repeating variables H_d (heat of combustion of the dust), P_{oe} (initial pressure in the enclosure), θ_{oe} (initial temperature in the enclosure), and V_e (volume of the enclosure). One or more of these repeating variables therefore appear in most of the Pi terms, except those which were already dimensionless.

Some of the terms in Table 4-2 can be converted to others which are more easily recognized. For example, form

$$\pi_{15}' = \frac{1}{\pi_{15}} \frac{1}{\pi_{16}} \frac{1}{\pi_{17}} = \frac{u' \ell_b}{\nu_b}$$

This is the Reynolds Number (N_{Re}) for turbulent eddies. It is also probably clearer to change most of the scaled density terms, to present them as density ratios. Some of the enclosure thermal response terms are converted, as follows:

$$\pi_{30}' = \frac{\pi_{30}}{\pi_{33}} = \frac{k_s}{\rho_s c_p V_e^{1/3}} = \frac{\text{heat conducted through enclosure}}{\text{heat stored in enclosure material}}$$

$$\pi_{32}' = \frac{\pi_{32}}{\pi_3} \cdot \pi_{31}^3 = \frac{\sigma V_e^{1/3} \theta_s^3}{k_s} = \frac{\text{heat radiated to enclosure}}{\text{heat conducted through enclosure}}$$

$$= \text{Fourier number } (N_{Fo})$$

TABLE 4-1 List of Parameters for Hybrid Explosions

Parameters	Symbol	Dimensions
Dust		
Mass	M_d	M
Heat of combustion	H_d	L^2/T^2
Mean particle size	d_d	L
Particle size distribution function	f_d	-
Moisture content (mass ratio)	R_w	-
Heat of vaporization of moisture (internal energy)	U_{fg}	L^2/T^2
Dust concentration	R_d	M/L^3
Average density of material in dust particles	ρ_d	M/L^3
Gas Mixture		
Heats of combustion of specific gas species	H_{ig}	L^2/T^2
Volume ratio of each gas species relative to air (stoichiometry)	ϕ_{ig}	-
Volume of each gas species at STP	V_{ig}	L^3
Ratio of specific heats for each gas	γ_i	-
Mean density of unburned gas	ρ_u	M/L^3
Burned Gas/Dust		
Laminar burning velocity	S_L	L/T
Turbulent burning velocity	S_T	L/T
Kinematic viscosity of combustion products	ν_b	L^2/T
Eddy size	ℓ_b	L
Mean flow velocity in turbulent eddies	u'	L/T
Mean density of combustion products	ρ_b	M/L^3
Ratio of specific heats for combustion products	γ_b	-
Temperature of combustion products	θ_b	θ
Enclosure & Initial Conditions		
Volume	V_e	L^3
Geometry	ℓ_{ei}	L
Initial pressure	p_{oe}	M/T^2L
Initial temperature	θ_{oe}	θ
Ratio of specific heats for air	γ_e	-
Ignition Source		
Energy	E_i	ML^2/T^2
Duration	t_i	T
Geometry	ℓ_{ij}	L
Temperature	θ_i	θ

TABLE 4-1 (Continued)

Parameters	Symbol	Dimensions
Pressure Response		
Pressure (a function of time)	p	M/T^2L
Rate of pressure change	$dp/dt = \dot{p}$	M/T^3L
Enclosure Thermal Response		
Thermal (radiative) absorptivity	α_{si}	-
Thermal conductivity	k_s	$ML/T^3\theta$
Initial temperature	θ_s	θ
Stefan-Boltzmann constant	σ	$M/T^3\theta^4$
Volumetric heat capacity	$\rho_s c_p$	$M/LT^2\theta$
Emissivity	ε_s	-
External ambient temperature	θ_o	θ
Time	t	T

TABLE 4-2. Pi Terms for Hybrid Explosions

NO.	SYMBOL	TERM	NO.	SYMBOL	TERM
Dust			**Enclosure and Initial Conditions**		
π_1	$\overline{M}_d$	$M_d H_d / p_{oe} V_e$	π_{21}	ℓ_{ei}	$\ell_{ei} / V_e^{1/3}$
π_2	$\overline{d}_d$	$d_d / V_e^{1/3}$	π_{22}	γ_e	γ_e
π_3	f_d	f_d	**Ignition Source**		
π_4	R_w	R_w	π_{23}	$\overline{E}_i$	$E_i / p_{oe} V_e$
π_5	$\overline{U}_{fg}$	U_{fg} / H_d	π_{24}	$\overline{t}_i$	$t_i H_d^{1/2} / V_e^{1/3}$
π_6	$\overline{R}_d$	$R_d H_d / p_{oe}$	π_{25}	ℓ_{ij}	$\ell_{ij} / V_e^{1/3}$
π_7	$\overline{\rho}_d$	$\rho_d H_d / p_{oe}$	π_{26}	$\overline{\theta}_i$	θ_i / θ_{oe}
Gas Mixture			**Pressure Response**		
π_8	$\overline{H}_{ig}$	H_{ig} / H_d	π_{27}	$\overline{p}$	p / p_{oe}
π_9	ϕ_{ig}	ϕ_{ig}	π_{28}	$\overline{\dot{p}}$	$\dot{p} V_e^{1/3} / p_{oe} H_d^{1/2}$
π_{10}	$\overline{V}_{ig}$	V_{ig} / V_e	**Pressure Response**		
π_{11}	γ_i	γ_i	π_{29}	α_{si}	α_{si}
π_{12}	$\overline{\rho}_u$	$\rho_u H_d / p_{oe}$	π_{30}	$\overline{k}_s$	$k_s \theta_{oe} / p_{oe} H_d^{1/2} V_e^{1/3}$
Burned Gas/Dust			π_{31}	$\overline{\theta}_s$	θ_s / θ_{oe}
π_{13}	$\overline{S}_L$	$S_L / H_d^{1/2}$	π_{32}	$\overline{\sigma}$	$\sigma \theta_{oe}^4 / p_{oe} H_d^{1/2}$
π_{14}	$\overline{S}_T$	$S_T / H_d^{1/2}$	π_{33}	$\overline{\rho_s c_p}$	$\rho_s c_p \theta_{oe} / p_{oe}$
π_{15}	$\overline{\nu}_b$	$\nu_b / H_d^{1/2} V_e^{1/3}$	π_{34}	ε_s	ε_s
π_{16}	ℓ_b	$\ell_b / V_e^{1/3}$	π_{35}	$\overline{\theta}_o$	θ_o / θ_{oe}
π_{17}	$\overline{u}'$	$u' / H_d^{1/2}$	π_{36}	$\overline{t}$	$t H_d^{1/2} / V_e^{1/3}$
π_{18}	$\overline{\rho}_b$	$\rho_b H_d / p_{oe}$			
π_{19}	γ_b	γ_b			
π_{20}	$\overline{\theta}_b$	θ_b / θ_{oe}			

With these conversions we obtain the modified set of the 36 Pi terms in Table 4-3.

The large number of terms in the similitude analysis may seem rather daunting. But, many are automatically satisfied if we retain the same fuels, same geometry and same initial conditions. Referring to Table 4-3, we see that term π_1 then requires that the dust concentration be the same, i.e., total mass of dust is proportional to enclosure volumes, Terms π_4, π_5, π_6' and π_7 are satisfied.

For the gas mixture, use of the same gases and some stoichiometry satisfies terms π_8 through π_{12}'.

For the combustion products, temperatures and ratios of specific heats should be the same (π_{13} and π_{20}), and so should all velocities (π_{13}, π_{14}, and π_{17}). Terms π_{21} and π_{22} are satisfied by the "replica" modeling we have assumed. The ignition source total energy should be scaled in proportion to the enclosure volume to satisfy π_{23}, and its duration should scale in proportion to the length scale factor from π_{24}. Terms π_{25} and π_{26} are satisfied if initial conditions and geometry are the same.

Term π_{27} tells us that pressures should not be changed by changing the length scale, and term π_{28} is essentially Bartknecht's cubic law for rate of pressure rise [Bartknecht (1978a)].

Replica scaling satisfies some of the terms for enclosure thermal response, namely π_{29}, π_{31}, π_{33}, π_{34}, and π_{35}. Term π_{36} is really general time scaling for replica model response, and it may or may not be satisfied for thermal model response of the enclosure.

Let us then present a reduced list of Pi terms which must concern us for the assumed replica modeling, by eliminating the ones we know to be satisfied. This list includes only seven terms, and appears in Table 4-4. Here, terms π_2 and π_3 tell us that the dust mean particle size should be scaled to the size of the enclosure, with the same dimensionless distribution function about the mean size. This may or may not be possible in practice.

Terms π_{15}' and π_{16} highlight a required compromise in the scaling. The Reynolds number term dictates the following relationship between scale factors:

$$\lambda_u' \cdot \lambda_{\ell_b} = \lambda_{v_b} \tag{1}$$

But, π_{16} requires that:

$$\lambda_{\ell_b} = \lambda_{V_e}^{1/3} = \lambda \tag{2}$$

where λ is the geometric scale factor.

TABLE 4-3. Modified Pi Terms for Hybrid Explosions

NO.	SYMBOL	TERM	NO.	SYMBOL	TERM
Dust			**Enclosure and Initial Conditions**		
π_1	$\overline{M}_d$	$M_dH_d/p_{oe}V_e$	π_{21}	$\bar{\ell}_{ei}$	$\ell_{ei}/V_e^{1/3}$
π_2	$\bar{d}_d$	$d_d/V_e^{1/3}$	π_{22}	γ_e	γ_e
π_3	f_d	f_d	**Ignition Source**		
π_4	R_w	R_w	π_{23}	$\overline{E}_i$	$E_i/p_{oe}V_e$
π_5	$\overline{U}_{fg}$	U_{fg}/H_d	π_{24}	$\bar{t}_i$	$t_iH_d^{1/2}/V_e^{1/3}$
π_6'	$\overline{R}_d$	R_d/ρ_d	π_{25}	ℓ_{ij}	$\ell_{ij}/V_e^{1/3}$
π_7	$\bar{\rho}_d$	ρ_dH_d/p_{oe}	π_{26}	$\bar{\theta}_i$	θ_i/θ_{oe}
Gas Mixture			**Pressure Response**		
π_8	$\overline{H}_{ig}$	H_{ig}/H_d	π_{27}	$\bar{p}$	p/p_{oe}
π_9	ϕ_{ig}	ϕ_{ig}	π_{28}	$\bar{\dot{p}}$	$\dot{p}V_e^{1/3}/p_{oe}H_d^{1/2}$
π_{10}	$\overline{V}_{ig}$	V_{ig}/V_e	**Pressure Response**		
π_{11}	γ_i	γ_i	π_{29}	α_{si}	α_{si}
π_{12}'	$\bar{\rho}_u$	ρ_u/ρ_d	π_{30}'	$\bar{k}_s'$	$k_s/\rho_sc_pV_e^{1/3}$
Burned Gas/Dust			π_{31}	$\bar{\theta}_s$	θ_s/θ_{oe}
π_{13}	$\overline{S}_L$	$S_L/H_d^{1/2}$	π_{32}'	N_{Fo}	$\sigma V_e^{1/3}\theta_s^3/k_s$
π_{14}	$\overline{S}_T$	$S_T/H_d^{1/2}$	π_{33}	$\overline{\rho_sc_p}$	$\rho_sc_p\theta_{oe}/p_{oe}$
π_{15}'	N_{Re}	$u'\ell_b/\nu_b$	π_{34}	ε_s	ε_s
π_{16}	ℓ_b	$\ell_b/V_e^{1/3}$	π_{35}	$\bar{\theta}_o$	θ_o/θ_{oe}
π_{17}	$\bar{u}'$	$u'/H_d^{1/2}$	π_{36}	$\bar{t}$	$tH_d^{1/2}/V_e^{1/3}$
π_{18}'	$\bar{\rho}_b$	ρ_b/ρ_d			
π_{19}	γ_b	γ_b			
π_{20}	$\bar{\theta}_b$	θ_b/θ_{oe}			

TABLE 4-4. Reduced List of Pi Terms for Replica Modeling

NO.	SYMBOL	TERM
Dust		
π_2	$\bar{d}_d$	$d_c/V_e^{1/3}$
π_3	f_d	f_d
Burned Gas/Dust		
π_{15}'	N_{Re}	$u'\ell_b/\nu_b$
π_{16}	ℓ_b	$\ell_b/N_e^{1/3}$
Enclosure Thermal Response		
π_{30}'	$\bar{k}_s'$	$k_s/\rho_s c_p V_e^{1/3}$
π_{32}'	N_{Fo}	$\sigma V_e^{1/3}\theta_s^3/k_s$
π_{36}	$\bar{t}$	$tH_d^{1/2}/V_e^{1/3}$

the replica scaling,

$$\lambda_u{}' = \lambda_{v_b} = 1 \tag{3}$$

When we substitute (2) and (3) into (1), we see that $\pi_{15}{}'$ must be distorted, if indeed the turbulent eddy size ℓ_b is scaled down in real life. But, if instead the turbulent eddies are the same size, then $\lambda_{\ell_b} = 1$, and term $\pi_{15}{}'$ <u>can</u> be satisfied.

It may be possible to satisfy terms $\pi_{30}{}'$ and $\pi_{32}{}'$ by some rather exotic means. The problems in simultaneous scaling of conductive and radiative heating of solid materials are addressed at some length in Chapter 12 of Baker, et al (1973). Practically, it probably is not worth the effort, because the effect of distortions in these heat transfer processes on the maximum gas temperatures and pressures and their rise rates, which are our primary concern, is minimal because the much slower thermal response. So, we should probably simply ignore these two Pi terms.

C. <u>Test methods</u>

Many of the closed vessel or closed-tube types of apparatus described in Chapters II and III for dust and gas explosion testing can also be used for testing of hybrid dust-gas explosion sources. The vertical detonation tube shown in Fig. 2-20 has been used for methane-air mixtures with added coal dust to demonstrate detonability of this hybrid explosion source [Lee and Guirao (1982)]. Peraldi and Veyssiere (1986) report using a similar apparatus to study explosions in hybrid mixtures of hydrogen-oxygen and ethylene-oxygen with added starch particles; and Veyssiere (1986) used the same apparatus to study detonations, in hybrid mixtures with aluminum particles suspended in mixtures of the fuels hydrogen, ethylene or acetylene with oxygen and inert gas nitrogen or argon. In both of these studies, distinctive two-peak detonation pressure records occurred, as shown in Fig. 4-1. The first peak corresponds to gaseous detonation and the second to the delayed contribution of reaction of the aluminum particles.

Campos, et al (1986), used another type of vertical tube apparatus to study effects of graphite added to lean methane-air mixtures in confined deflagrations. Their equipment is illustrated in Fig. 4-2. In tests in this apparatus, the authors reported no reaction from the graphite dust in any of their experiments.

Gaug, et al (1986) report tests to determine lean flammability limits of hybrid mixtures in two sizes of cylindrical vessels, shown schematically in Fig. 4-3 and 4-4. They tested mixtures of hydrogen-methane-air, hydrogen-iron-air and hydrogen-cornstarch-air. As expected, addition of combustible dusts lowered the lean flammability limit, but they noted that lean limit dust flame thickness could approach the minimum vessel dimension for small vessels. So, they

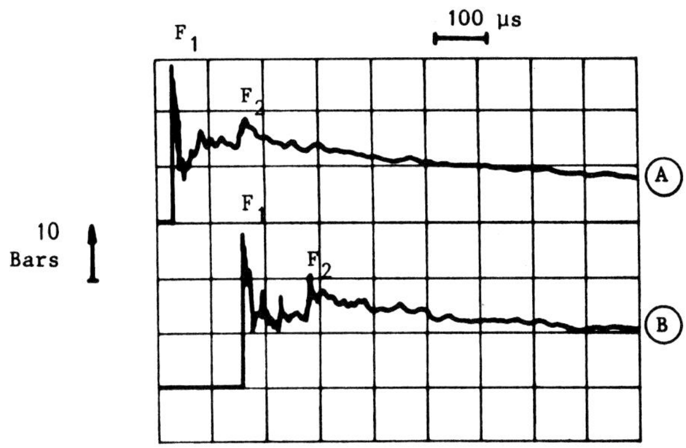

Fig. 4-1. Evolution of pressure in a mixture of combustible gas with aluminum particles: (A) at 1.945m from the ignition point; (B) at 4.175m from the ignition point.

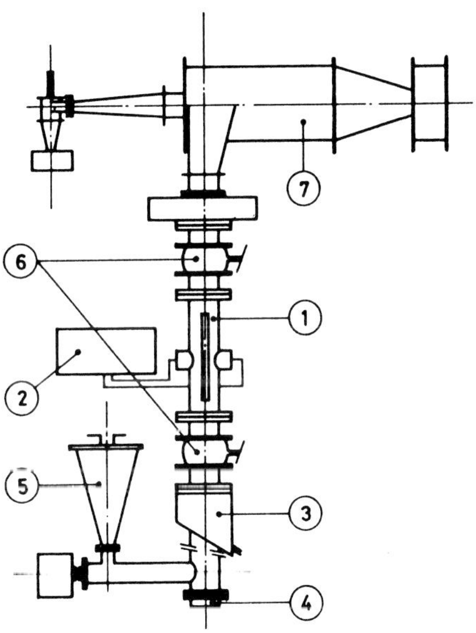

Fig. 4-2. Experimental apparatus of Campos, et al (1986): 1) combustion chamber; 2) spark ignition system; 3) dust suspension generator; 4) methane-air feed; 5) graphite feed; 6) valves; 7) cyclone.

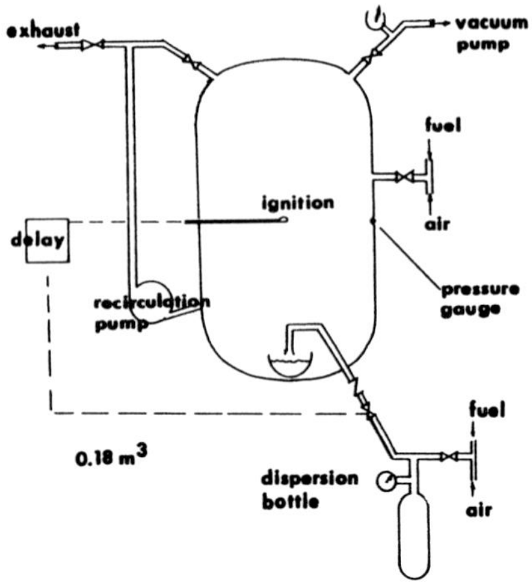

Fig. 4-3. 0.18 m^3 cylindrical dust bomb [Gaug, et al (1986)].

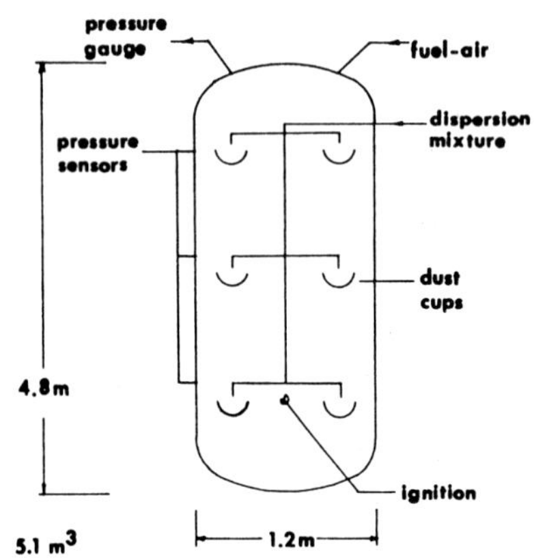

Fig. 4-4. 5 m^3 cylindrical apparatus [Gaug, et al (1986)].

recommended use of large-scale apparatus whenever possible in determining such limits.

Klemens and Wolanski (1986) report the use of a vertical duct apparatus, open at the bottom, to study flame structure in hybrid mixtures near lean flammability limits. They tested methane-air-coal and methane-air-lignite mixtures. The gas-air mixture was passed through the duct for several minutes before dust dispersion and ignition. Again, they demonstrated reduction of the lean flammability limit. This apparatus and other types of vertical apparatus usually allow good optical coverage of the flame on detonation fronts, as well as pressure and/or temperature sensing.

Bradley, et al (1984) describe but do not illustrate apparatus for observing and measuring the effects on flame fronts in gas flames of adding inert (alumina) and reactive (graphite) dusts. These authors report that, for relatively small dust concentrations, the added dust simply acts as a heat sink for the inert dust and as a heat source for the reactive dust. Only small changes in burning velocity were observed. But, as dust concentration increased, this simple assumption was no longer valid.

D. Control methods

Methods to control hybrid explosions are essentially the same as for confined gas or dust explosions, that is, venting, inerting, or suppression. The only variation we could find involved the dispersion of inert powders or dusts into flammable gas mixtures to quench flames near lean flammability limits. Similarly, in coal mines, rock dust is a common material used to control the methane-coal-air hybrid mixtures which are so dangerous in coal mines.

E. Closure

It is no accident that this is the shortest chapter in this book. The reader should understand, if he has read this far, that dust explosions and reactive gas explosions both involve very complex physical processes, and so are very difficult to analyze and involve difficult and expensive testing. So, combining these two types of explosion sources can only lead to further complexity. This is reflected by the paucity of test data for hybrid explosions and the almost complete lack of analyses.

All one can state with confidence regarding hybrid explosions is that some such explosion sources, particularly in coal mines, can be more easily ignited than either gas or dust explosions alone, and that they can sometimes cause much more damage in confined spaces such as mines.

V. NON-REACTIVE GAS AND FLASH-EVAPORATING FLUID EXPLOSIONS

A. Fundamentals

The class of explosion covered in this chapter is sometimes referred to as a simple pressure burst. No chemical reactions occur, the cause of the explosion being a sudden failure of a pressure vessel or piping system. Some such failures can occur at normal operating pressures in a flawed vessel or pipe, when a subcritical flaw slowly grows due to stresses, vibration or perhaps corrosion in service, and then reaches a critical size which results in a rapid vessel failure. The stored energy in the compressed fluid in the vessel serves both as the agent which accelerates the vessel failure and as the explosion source. This type of explosion is most dangerous for compressed gases, next most dangerous for compressed flash-evaporating fluids, and least dangerous for any compressed liquid. Because blast waves are inconsequential for compressed liquid vessel failures, we discuss in this chapter only explosions releasing compressed gases and flash-evaporating fluids.

Much of the basic analytic and experimental work related to this class of explosion has been sponsored by NASA and is reported in NASA reports [Baker, et al (1975), Strehlow and Ricker (1976), Baker, et al (1978), Esparza and Baker (1977a) Adamczyk and Strehlow (1977), and Esparza and Baker (1977b)]. Brode (1955) did significant early analytic work on bursting, gas-pressured spheres, and Boyer, et al (1958) conducted some experiments using optical instrumentation only. Pittman (1972) and (1976) has also reported limited test data for metal spherical vessels pressured to failure, and Baum (1979) reports limited blast measurements for bursting pipe sections. These references are the basis for the development in the remainder of this chapter.

There are some basic differences in the properties of air blast waves developed by vessel bursts and those developed by detonation of condensed-phase high explosives such as TNT. An idealized profile of a wave from a typical TNT explosion is shown in Fig. 5-1. The rise to peak shock overpressure P_s is essentially instantaneous at shock front arrival, and pressure decay is initially quite rapid, falling below atmospheric pressure and finally recovering to ambient more slowly. A secondary shock is often evident, but its amplitude is always small compared to the primary shock. The secondary shock is caused by a rarefaction wave which propagates back into the explosion products as the primary shock is transmitted to surrounding air. On reaching the center, the rarefaction reflects as the secondary shock wave. The primary wave is usually characterized by its amplitude P_s, positive phase duration t_d, and positive phase specific impulse i_s. Negative phase parameters have seldom been reported or computed.

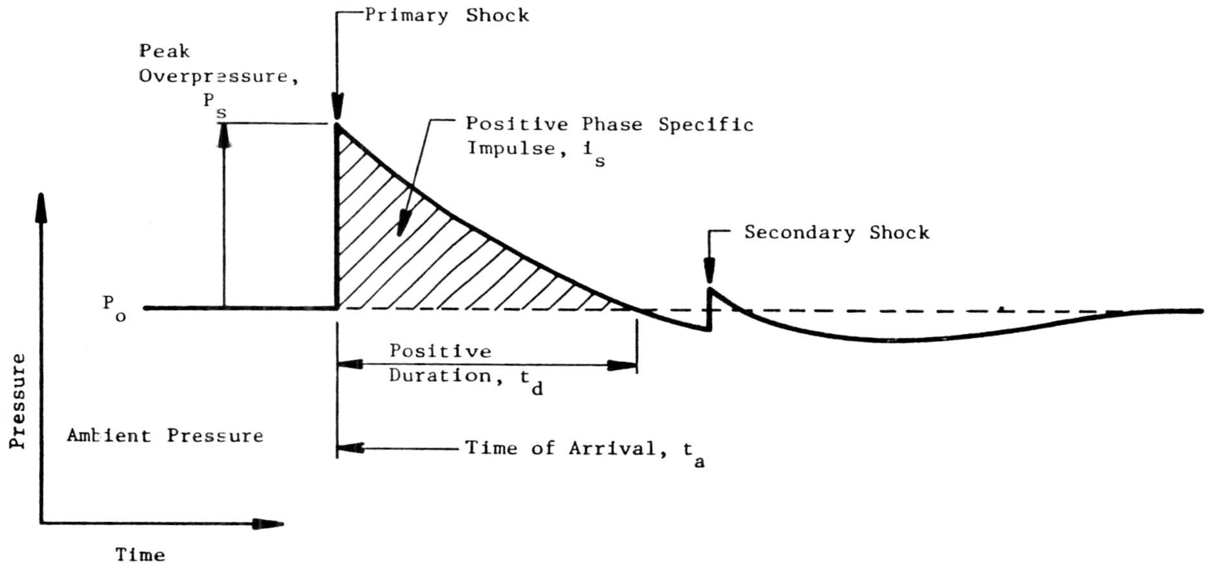

Fig. 5-1. Idealized profile of a blast wave from a condensed high explosive.

Quite complete prediction curves and equations for these and other free-field and reflected blast wave properties, with the data based being primarily tests with TNT or Pentolite explosives, appear in Baker (1983), Baker, et al (1983), and other compendia of blast data. We will show some of these curves later when comparing predictions of blast wave properties for various sources.

Blast waves from gas vessel bursts show some similarity to those from condensed explosives, but their lower energy densities and larger sources required for equivalent total energies also change the wave characteristics. An idealized profile for such bursts is shown in Fig. 5-2. The negative phase is much more pronounced than for TNT explosions, and the second shock is much stronger, compared to the first shock, because the rarefaction wave is more pronounced. There are some data available for all the first positive phase, negative phase, and second shock properties indicated in Fig. 5-2. We will show later that blast waves for bursts of vessels containing flash-evaporating fluids have characteristics quite similar to the waves for bursts of vessels containing compressed gases.

B. Scaling

In essence, the scaling law for blast waves from bursting pressure systems is a slightly reduced version of the law already stated in Chapter III for blast from reactive gas explosions. It was developed in Esparza and Baker (1977a), and is simply stated here without proof. Using the same format as in Chapter III, this law can be stated as:

$$\left.\begin{aligned} \bar{p} &= (p/p_o) \\ \bar{t}_a &= (t_a a_o p_o^{1/3}/E^{1/3}) \\ \bar{t}_d &= (t_d a_o p_o^{1/3}/E^{1/3}) \\ \bar{i}_s &= (i_s a_o/p_o^{2/3}E^{1/3}) \end{aligned}\right\} \begin{aligned} &= f_i\left[\frac{R p_o^{1/3}}{E^{1/3}}, \left(\frac{p_1}{p_o}\right), \gamma_1, \frac{a_1}{a_o}, \ell_i\right] \\ &= f_i(\bar{R}, \bar{p}_1, \gamma_1, \bar{a}_1, \ell_i) \end{aligned} \tag{1}$$

Note that only the scaled energy release rate has been dropped from the earlier statement, because this rate is undefined for a pressure burst. All terms are defined in Table 3-3 in Chapter III.

This law will be used throughout this chapter for concise presentation of analytic and experimental results.

C. Results of Analyses

As noted earlier, a number of authors have analytically predicted the characteristics of blast waves from bursting pressure vessels. Most authors, starting with Brode (1955), (1956) and Boyer, et al (1958), have used one-dimensional hydrocodes for such analyses. They have usually concentrated on defining free-field blast wave properties as a function of vessel burst conditions and distance R from the center of the vessel. Furthermore, most analysts have limited their calculations to properties of the first positive phase of the blast wave, rather than the more complete set of properties indicated in Fig. 5-2. Baker, et al (1983) includes a rather complete discussion of systematic calculations of this kind made by Strehlow and Ricker (1976), and also presents smooth, scaled curves to allow prediction of the first positive phase overpressure P_{s1} and specific impulse i_{s1}. These curves are included here as Figs. 5-3 and 5-4.

In support of a NASA handbook [Baker, et al (1978)], Ricker extended his previous hydrocode calculations to include the two-dimensional blast wave properties for a spherical pressure vessel bursting into halves, as shown in Fig. 5-5. Again, he concentrated on positive phase properties. Predicted blast properties were, of course, directional, being strongest along the plane of separation of the halves. Scaled curves for first positive phase overpressure are given here in Fig. 5-6. Impulse was found to be essentially non-directional, and can be predicted from Fig. 5-4.

Some supporting equations are needed to use Figs. 5-3, 5-4, and 5-6 for blast wave predictions. The shock tube equation can be used to calculate an initial dimensionless pressure at the sphere surface to enter Figs. 5-3 or 5-6, and a fictitious dimensionless sphere radius $\bar{R}_1$, calculated from the enclosure volume, burst pressure and γ_1. The applicable equations are [Baker, et al (1975)]

$$\frac{p_1}{p_o} = \frac{p_{so}}{p_o} \left\{ 1 - \frac{(\gamma_1 - 1)(a_o / a_1)\left(\frac{p_{so}}{p_o} - 1\right)}{\sqrt{(2\gamma_o)\left[2\gamma_o + (\gamma_o + 1)\left(\frac{p_{so}}{p_o} - 1\right)\right]}} \right\}^{\left(\frac{-2\gamma_1}{\gamma_1 - 1}\right)} \tag{2}$$

$$E = \frac{p_1 - p_o}{\gamma_1 - 1} V_1 = \frac{4\pi}{3} \frac{p_1 - p_o}{\gamma_1 - 1} r_1^{\,3} \tag{3}$$

$$\bar{R}_1 = \left[\frac{3(\gamma_1 - 1)}{4\pi\left(\frac{p_1}{p_o} - 1\right)} \right]^{1/3} \tag{4}$$

$$\bar{R}_1 = \frac{r_1 p_o^{1/3}}{E^{1/3}} \tag{5}$$

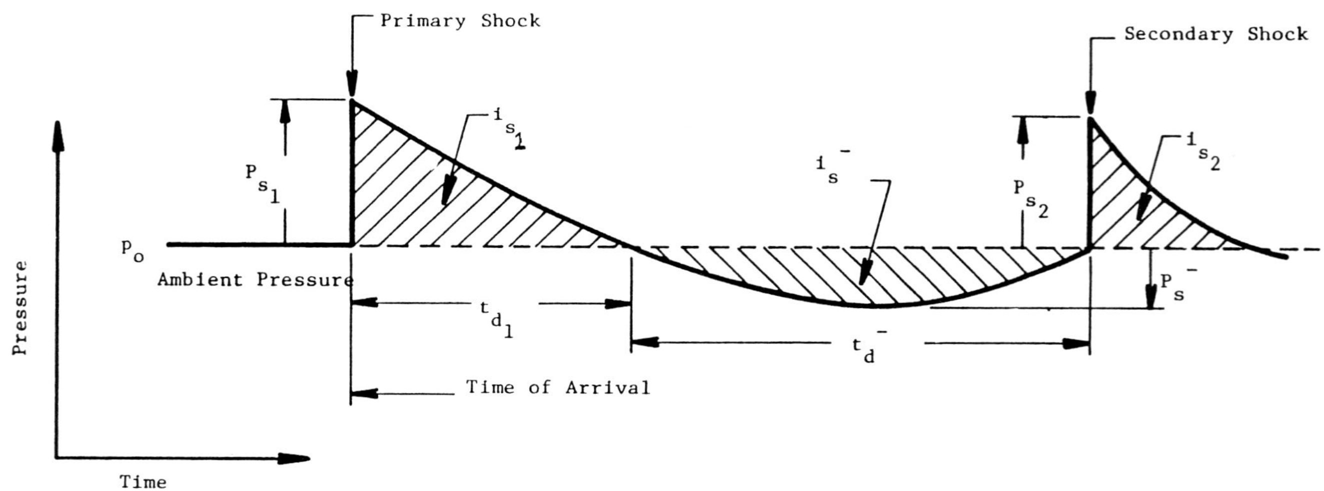

Fig. 5-2. Idealized profile of a blast wave from a bursting gas vessel.

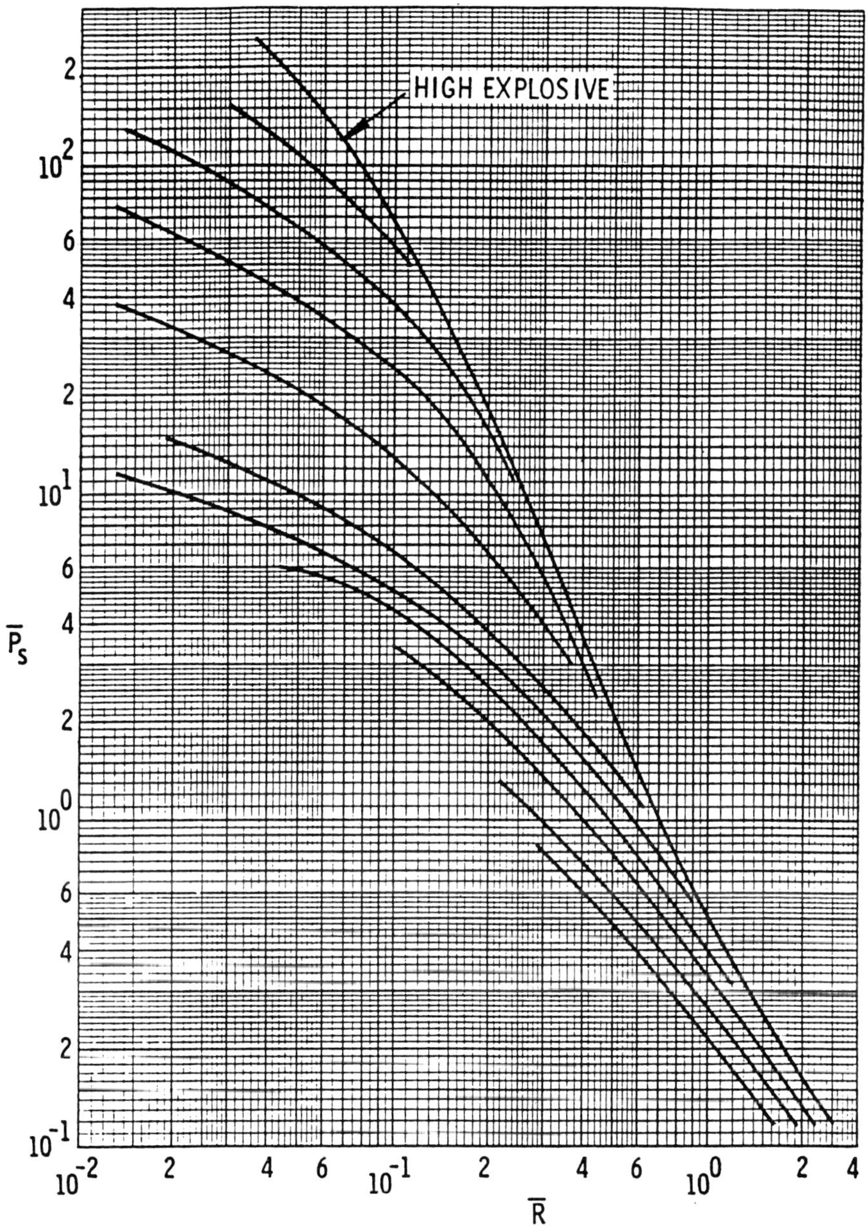

Fig. 5-3. Dimensionless shock overpressure versus energy-scaled distance, spherically symmetric vessel bursts. [Baker, et al (1983)]

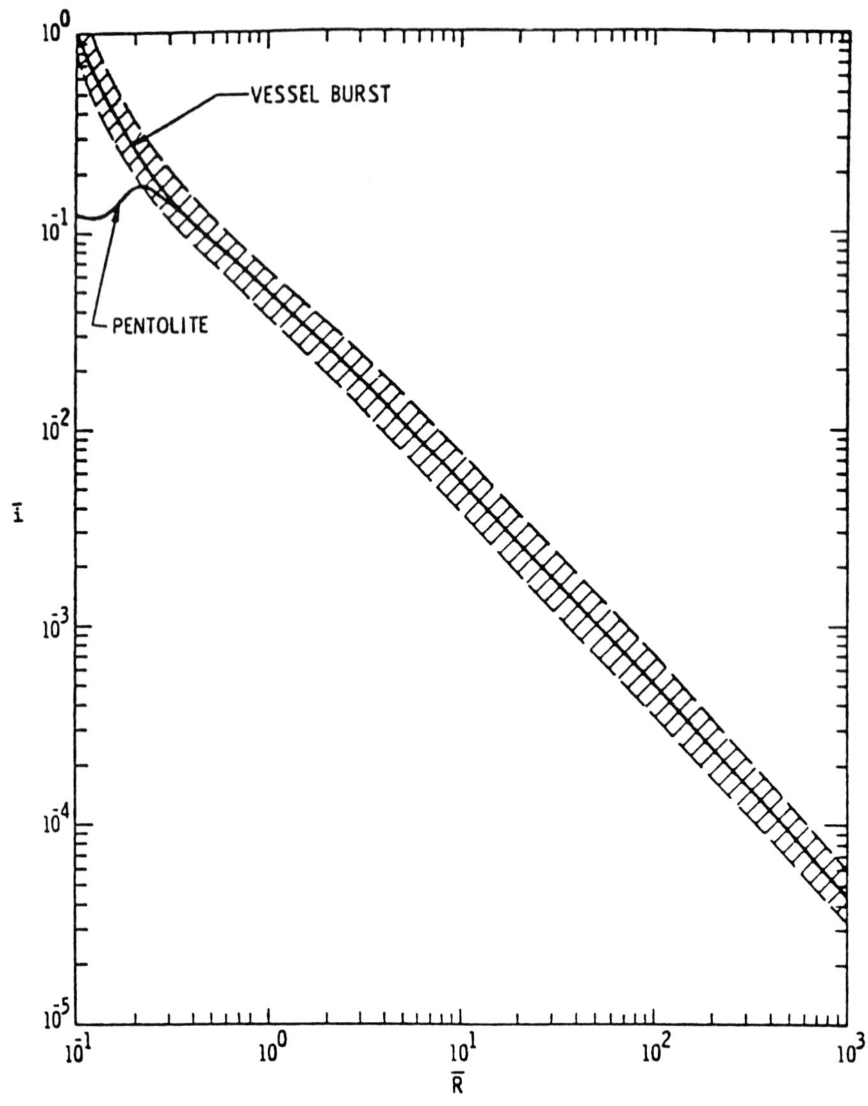

Fig. 5-4. Dimensionless positive phase specific impulses versus energy-scaled distance, all vessel bursts. [Baker, et al (1983)]

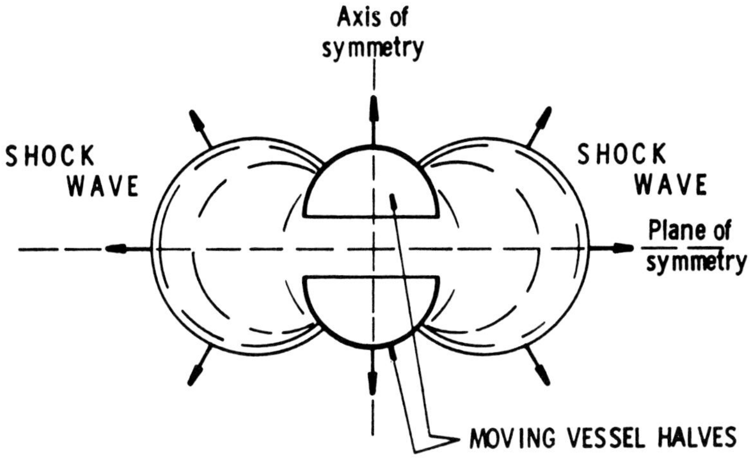

Fig. 5-5. Burst of a spherical pressure vessel. [Baker, et al (1978)]

In Eq. (2), p_{so}/p_o is the dimensionless air shock pressure at the instant of burst, p_1/p_o is the dimensionless sphere pressure, and a_o/a_1 is the velocity of sound ratio. Note that the sphere dimensionless shock overpressure is

$$\overline{P}_{so} = (p_{so}/p_o - 1)$$

Usually, it is the unknown in Eq. (2), and one must solve the equation by iteration to obtain p_{so}/p_o.

To "unscale" values from these graphs, one must invert three of the dimensionless parameters in Eq. (1), i.e.,[1]

$$P_s = \overline{P}_s p_o \tag{7}$$

$$i_s = \bar{i}_s \frac{p_o^{2/3} E^{1/3}}{a_o} \tag{8}$$

$$R = \overline{R} \left(\frac{E}{p_o} \right)^{1/3} \tag{9}$$

[1]Note: For explosions occurring on the ground surface, blast wave reflection is usually accounted for by multiplying by a reflection factor of 2 on E.

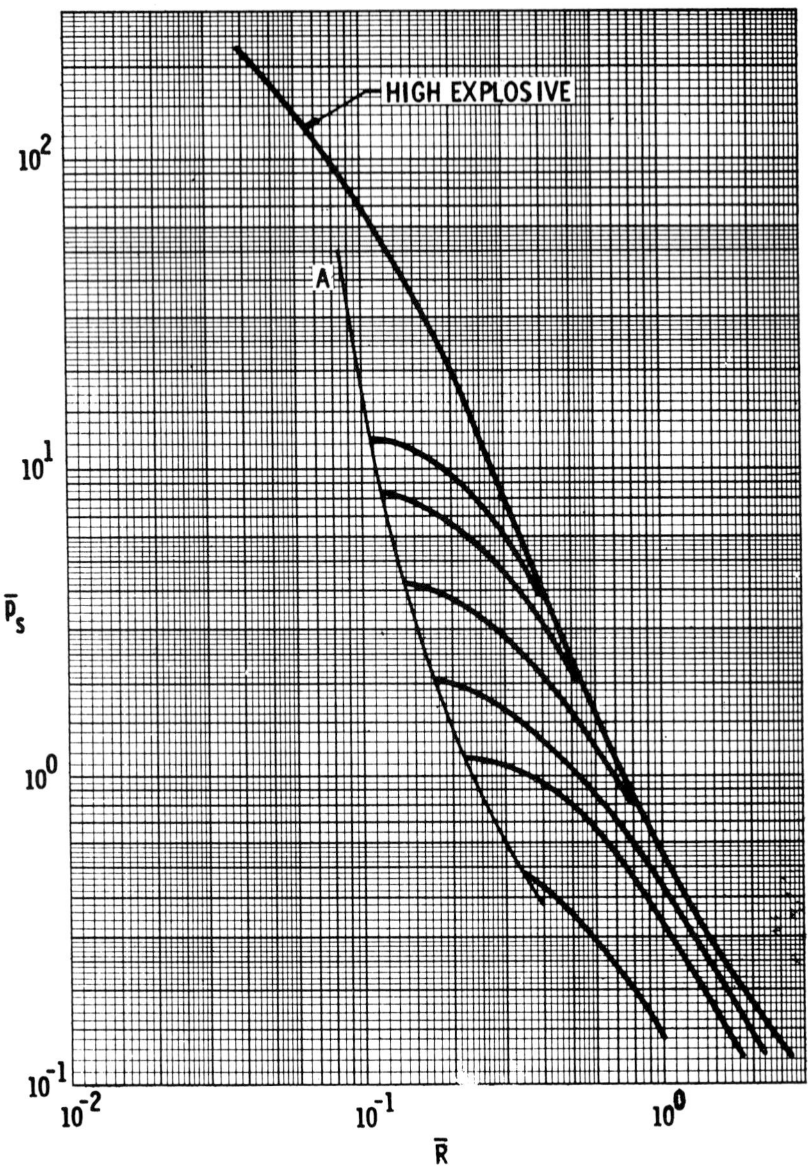

Fig. 5-6. Dimensionless peak shock overpressure versus energy-scaled distance, maxima for sphere bursting in half. [Baker, et al (1978)]

The quantity $(E/p_o)^{1/3}$ has dimensions of length, and it is called the characteristic length for the explosion source by some authors.

The ratios of specific heat, γ_o and γ_1 are dependent on the gases in the sphere (γ_1) and the surrounding atmosphere (γ_o). Normal choices are:

$$\gamma = 5/3 \text{ (monatomic gas)} \quad (6)$$

$$\gamma = 7/5 \text{ (diatomic gas)}$$

$$\gamma = 1.2 \text{(average for combustion products)} \quad (10)$$

Surrounding atmosphere is usually air, making $\gamma_o = 7/5 = 1.4$

Raju and Strehlow (1984) have used a hydrocode to calculate the directional characteristics of the blast wave from a bursting oblate spheroid vessel with aspect ratio (AR) of five. They simulated burst conditions for constant-volume combustion of a stoichiometric methane-air mixture in the spheroid of p_1/p_o = 8.934; γ_o = 1.4 and γ_1 = 1.202. Dimensionless source internal energy was

$$\overline{E} = e\frac{\rho_o}{\rho_o} = 44.24 \quad (11)$$

where ρ_o is initial gas density and e is initial internal energy per unit volume. Figs. 5-7 and 5-8 show graphic results of their analyses. Scaling is according to Eq. (1). Note that in these figures the long axis of the spheroid corresponds to the direction $\phi = 90^\circ$, and the short axis to $\phi = 0^\circ$. The authors note that the blast front is initially spheroidal in shape, but that it quickly approaches a spherical shape. Note that directional effects on peak overpressure and specific impulse seem to disappear at a scaled distance of R 1. As with some other calculations by Strehlow and his coworkers, the calculations predict approximate TNT equivalence based on peak overpressure at the larger scaled distances, but do not approach equivalence based on scaled impulse within the range of the analyses.

Unlike the calculations for spherical pressure bursts, the work of Raju and Strehlow for the two-dimensional burst is not extensive enough to allow blast wave prediction for a variety of vessel and gas burst conditions. It does, however, give some good insight into close-in shape effects for one non-spherical geometry.

For strong enclosures of large L/D ratios, such as piping or in some instances tanker vessels or buildings, the prediction curves for blast wave properties for spherical or small L/D may be inapplicable because the waves will be generated by essentially line sources. There have unfortunately been no systematic treatments of bursting vessel blast from such sources, either

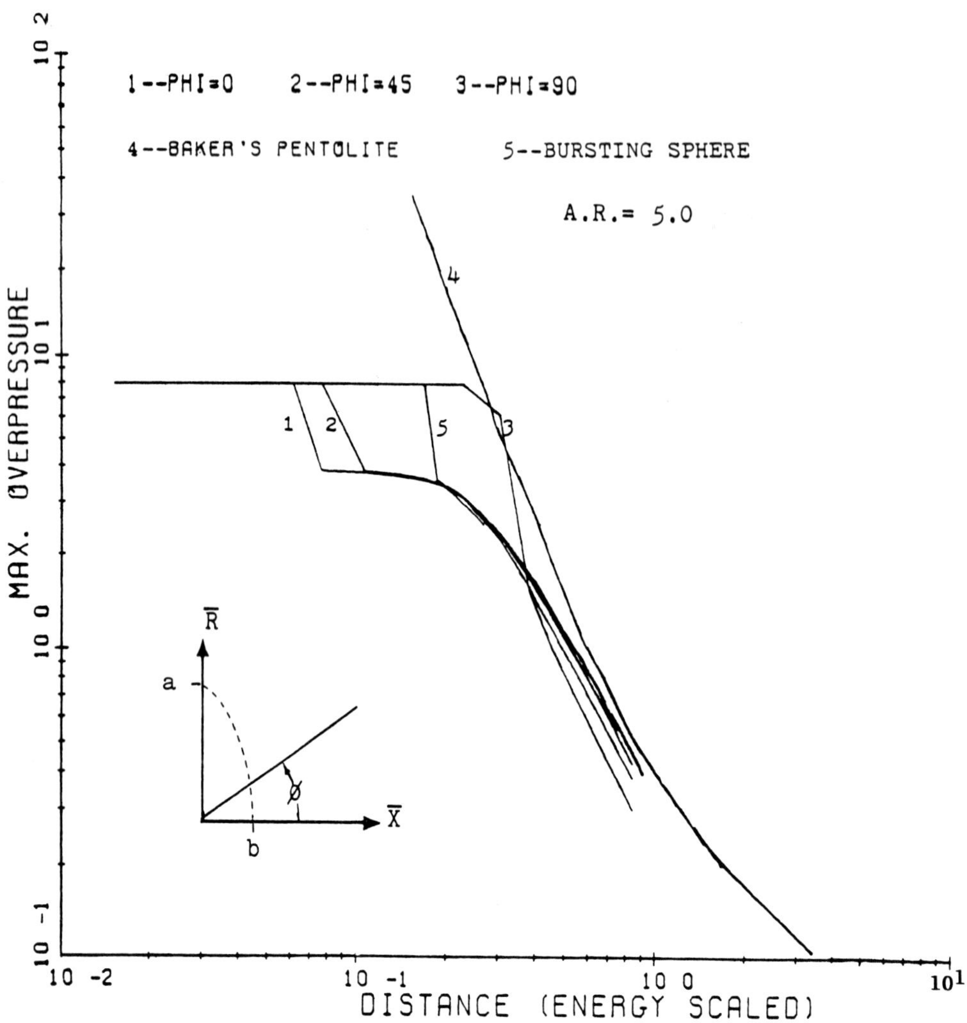

Fig. 5-7. Dimensionless air blast overpressure for burst of a spheroidal pressure vessel of aspect ratio = 5. [Raju and Strehlow (1984)]

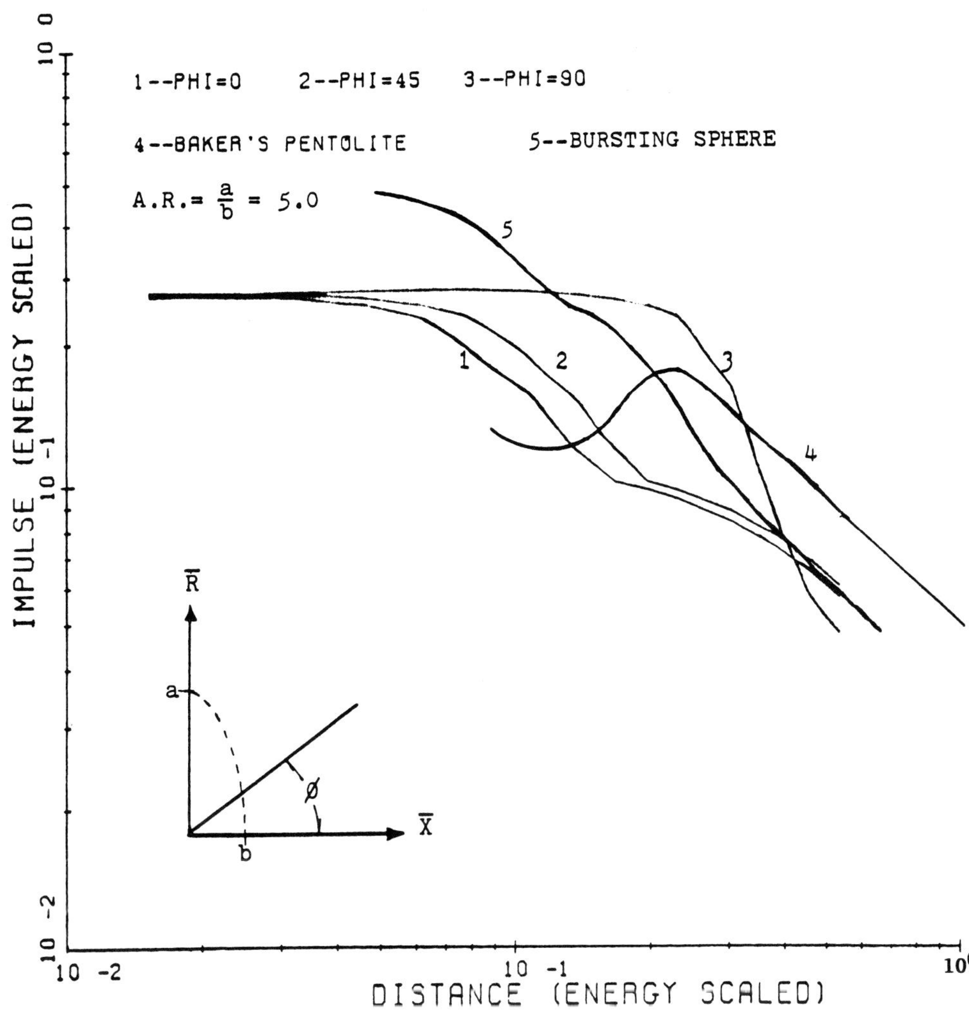

Fig. 5-8. Dimensionless specific impulse for burst of a spheroidal pressure vessel of aspect ratio = 5. [Raju and Strehlow (1984)]

analytically or experimentally, so no adequate predictions of the blast wave properties can be made, particularly close to the bursting enclosure. The only analysis the author has found is by Baum (1979), who has developed an approximate method for blast prediction close to longitudinal ruptures in gas pressurized pipes.

D. Results of tests

The first blast tests of bursting pressure vessels the author is aware of were conducted by Boyer, et al (1958). They pressurized glass spheres and ruptured them with a striker, and observed the resulting expanding shock fronts with spark shadowgraphs and schlieren photography. A schlieren picture of one of their tests is reproduced here in Fig. 5-9. But, these investigators made no measurements of blast wave properties other than shock front locations versus time.

Pittman (1972) later conducted a series of burst tests of small pressure tanks, and made limited measurements of the blast from these pressure bursts. Again, the measurements were too few to develop a significant data base.

More recently, Esparza and Baker (1977 a and b) conducted two series of small-scale experiments to obtain free field blast data at various distances from bursting pressurized spheres. Glass spheres of nominal 51 and 102 mm diameter under internal pressure were ruptured by a striker and complete time histories of overpressure were obtained with an array of side-on pressure transducers at different radial distances. The test arrangement is shown schematically in Fig. 5-10.

Each experiment was set up in the test cell as shown in Fig. 5-11. Once the sphere was properly connected and all instrumentation ready to record data, the pressuring system was purged several times and the sphere filled with the appropriate test fluid. After the temperature of the fluid stabilized to the ambient value, the solenoid valve in the filling line was closed remotely. The high-speed camera and the spotlight were then turned on to begin the actual test. At a preset point of film travel the contacts in the camera closed which energized the solenoid on the pneumatic cylinder. The cylinder was pressurized and the striker burst the sphere releasing the high-pressure fluid.

The output of the pressure transducers was recorded on Polaroid film using several digital transient recorders and oscilloscopes.

The reduced data from these tests were scaled using the scaling law of Eq. (1), and converted to graphic plots. In addition to the pressure measurement system, high-speed cinematography was used in some of the tests to observe sphere breakup and obtain velocities of glass fragments. The velocity data were used to obtain the energy driving the blast wave by computing the fragment kinetic energy and subtracting it from the initial energy in the compressed fluid.

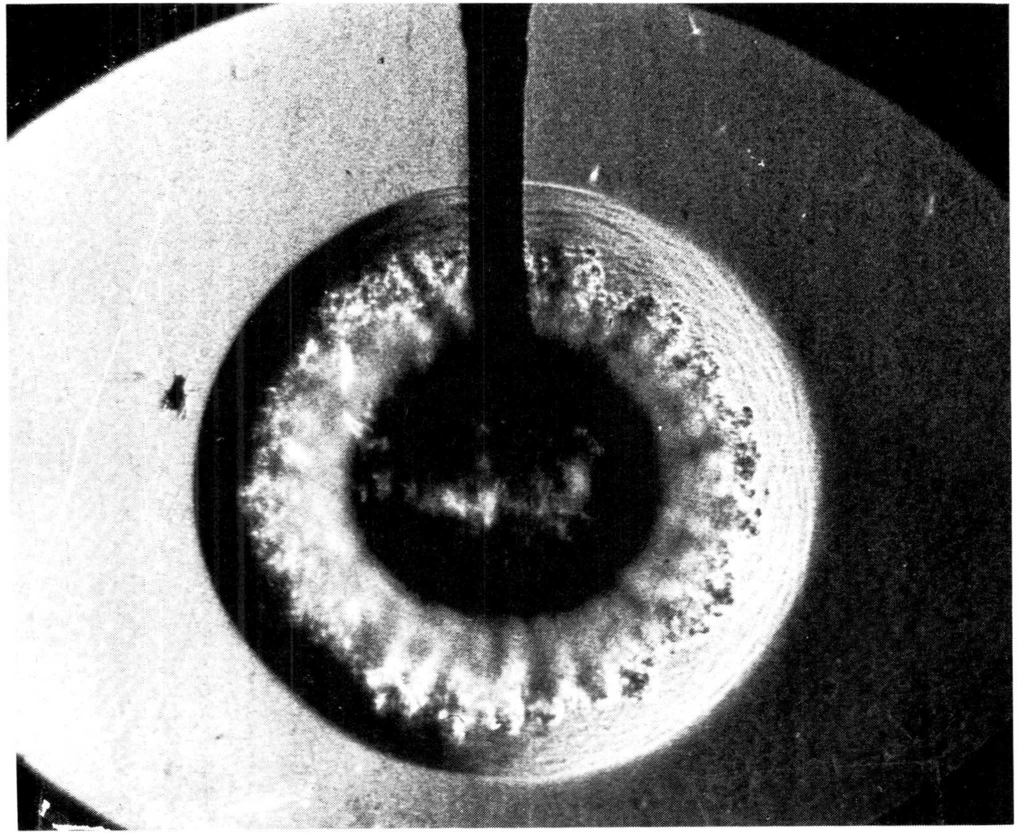

Fig. 5-9. Schlieren picture of blast from a pressurized glass sphere. (Courtesy of Dr. I. I. Glass, Univ. of Toronto)

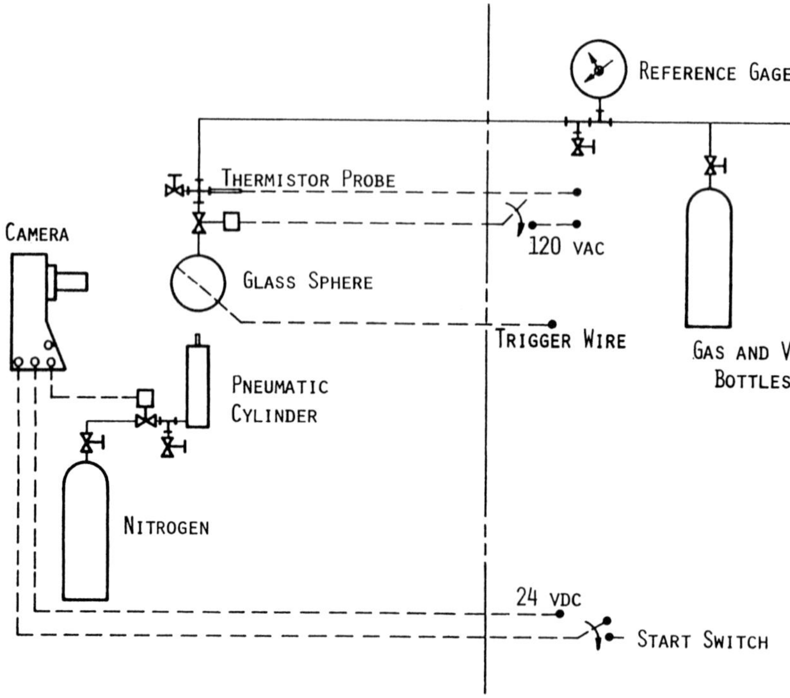

Fig. 5-10. Diagram of experimental set-up for glass sphere burst tests. [Esparza and Baker (1977a&b)]

Fig. 5-11. Actual test arrangement for glass sphere burst tests. [Esparza and Baker (1977a&b)]

An important parameter in these tests, as in most blast work, is of course the total source energy E. For the tests with pressurized gases, this value was calculated from Eq. (3), less about 10% for kinetic energy of the glass sphere fragments. For tests with the flash-evaporating Freon 12® fluid, source energy had to be calculated differently. Esparza and Baker (1977b) proposed calculation of the source energy by assuming isentropic expansion of the fluid from its initial state to atmospheric pressure, and calculating the change in internal energy u, and then multiplying by the mass of fluid. [This method is also described in Baker, et al, (1983).] In equation form, then, the internal energy change is

$$u_1 - u_2 = \int_1^2 p\,dv \tag{12}$$

and

$$E = m\,(u_1 - u_2) \tag{13}$$

where p is absolute pressure, v is specific volume, and m is mass of fluid. In practice, one obtains u_1 and u_2 from tables of thermodynamic properties of the expanding fluid assuming isentropic expansion from initial state conditions p_1, v_1 to atmospheric pressure p_2. Baker, et al (1983) gives an example problem for such a calculation.

In all of the gas burst tests, quite distinctive blast waves were recorded, similar to the idealized waveform in Fig. 5-2. A redrawn record appears in Fig. 5-12. This figure also indicates nomenclature for the additional parameters needed to describe this relatively complex waveform.

To compare with analyses of Strehlow and Ricker (1976) and with scaled Pentolite data of Baker (1973), the scaled test data for first shock overpressures and specific impulse were plotted versus dimensionless scaled distance $\bar{R}$, see Figs. 5-13 through 5-16.

From the graphic comparisons in Fig. 5-13, one can see that the analytic predictions for first positive phase air blast properties for gas bursts are conservative (too high) for $\bar{R} < 2$, but agree well with the scaled test data above that value. Scaled specific impulses for first positive phase (Fig. 5-14) agree quite well with compiled data for high explosives.

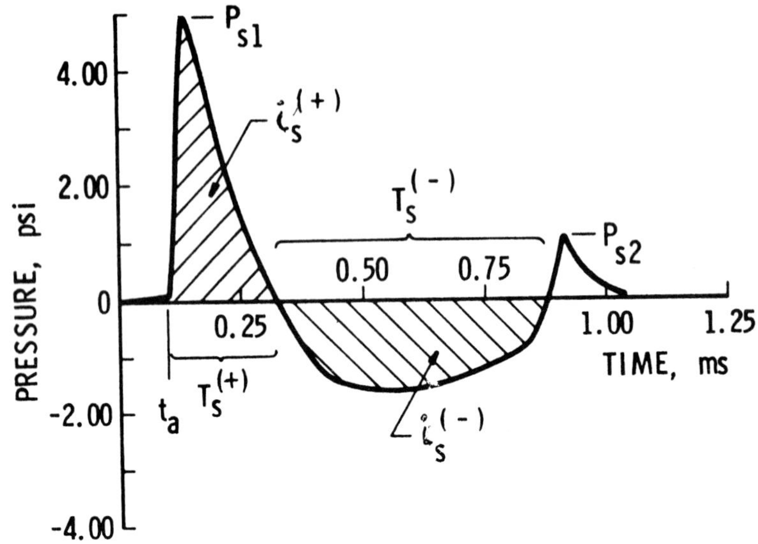

Fig. 5-12. Typical blast pressure history for frangible gas sphere burst. [Esparza and Baker (1977a)]

Typical pressure traces from Ref. 1 for bursting spheres filled with Freon-12® saturated vapor, shown in Fig. 5-16, indicate the same general characteristics as for gas-filled vessel, except that the first pressure pulse is much more rounded and the second shock often has greater amplitude than the first shock. Data for bursts with Freon 12® vapor show scaled first shock overpressures well below those of Pentolite high explosive (Fig. 5-16) while first positive phase specific impulses are slightly greater (Fig. 5-17). We have no analytic predictions to compare to these data.

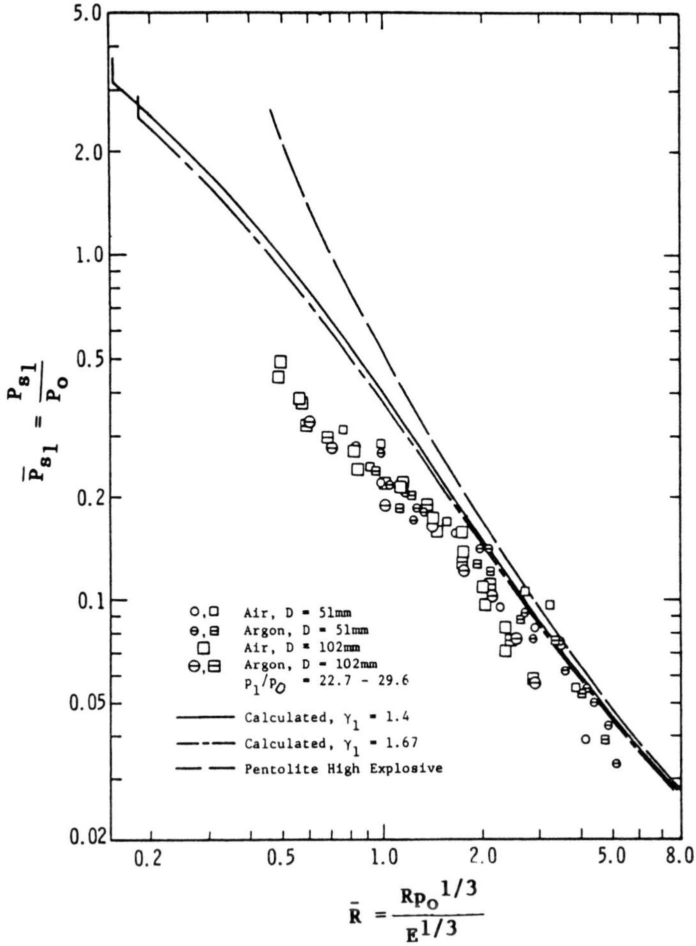

Fig. 5-13. Scaled side-on peak overpressure from bursting gas spheres. [Esparza and Baker (1977b)]

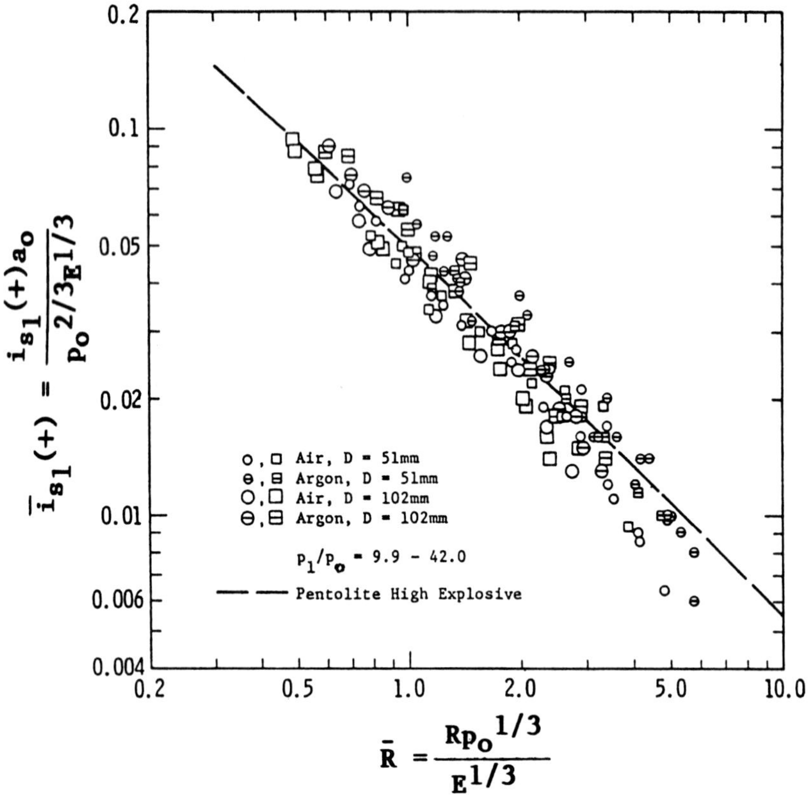

Fig. 5-14. Scaled side-on positive impulse from bursting gas spheres. [Esparza and Baker (1977a)]

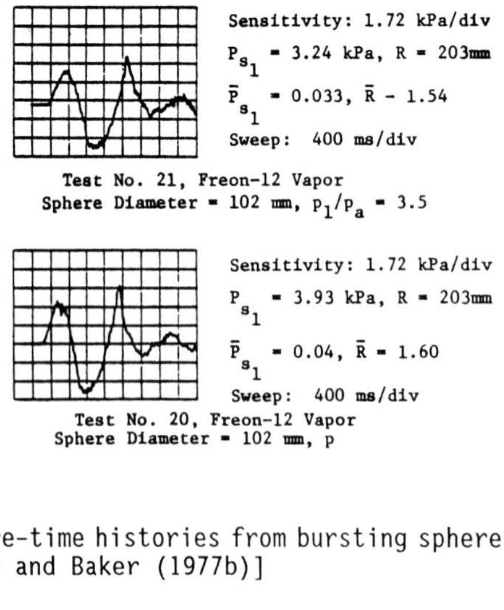

Fig. 5-15. Pressure-time histories from bursting spheres pressurized with Freon-12® vapor [Esparza and Baker (1977b)]

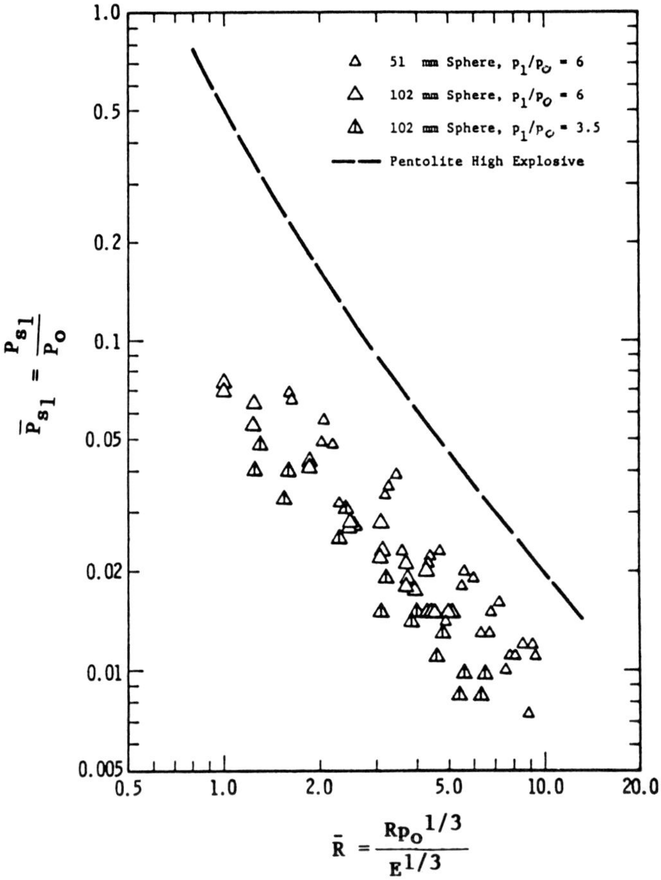

Fig. 5-16. Scaled side-on peak overpressure, Freon-12® vapor sphere at room temperature. [Esparza and Baker (1977b)]

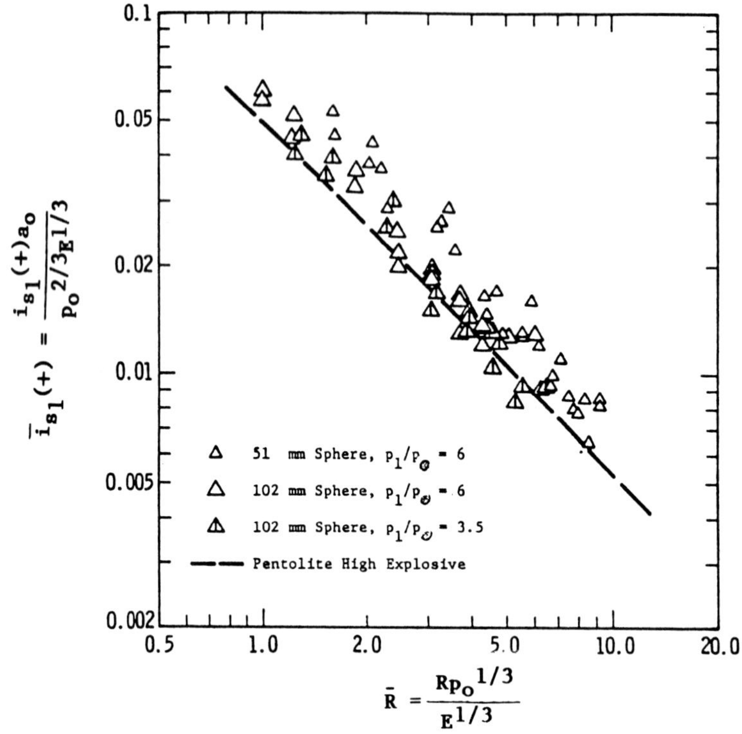

Fig. 5-17. Scaled side-on positive impulse from bursting Freon-12® vapor sphere. [Esparza and Baker (1977b)]

The scaling law of Equation (1) does not and cannot show what the functional forms f_1 are, nor does it tell the relative importance of varying each of the parameters. Either analysis or experiment or both must be conducted to get these answers. What it does do is to show a convenient way of presenting results of tests of analyses, or comparing results from various investigators. Ideally, one should vary each of the last four parameters in the bracket in Eq. (1) while holding the other three constant, and determine the scaled blast parameters as functions of scaled distance. However, for the testing reported here the scaling law was simplified for each type of blast source. Only air and argon were used as pressurized gases in the blast wave source. For air $\gamma_1 = 1.4$ and for argon $\gamma_1 = 1.667$, which is a minor difference. The pi term a_1/a_o equals one for air and very nearly one for argon. Therefore, this term can be eliminated. Finally, the previous theoretical calculations [Strehlow and Ricker (1976)] and the experimental data indicated that blast wave characteristics were only weakly dependent on the initial pressure ratio (p_1/p_o). This effective ratio was varied in these experiments over a range of 9.9 to 42.0. Consequently, all the blast data are combined for the two gases.

These simplifications let us plot the data strictly as functions of dimensionless scaled distance $\bar{R}$ (Sachs' scaled distance). This is done in Figs. 5-18 through 5-25.

The parameters presented are the times of arrival of the first and second shock, the peak overpressures of both shocks, the first positive and negative phase impulses, and the durations of these two phases. The figures show the range of all test data within the cross-hatched areas, and a "best fit" solid curve through the data.

Similarly, for the Freon 12® vapor tests, the blast data can be presented as a function of only scaled distance because the ratios of specific heats, sound velocities, and densities are invariant among these experiments, and the internal pressure ratios used were only 3.5 and 6.0. These data are presented in Figs. 5-26 through 5-33 as a function of scaled distance. Note that the data from the gas and vapor experiments, in general, will not plot together because each blast parameter will be a different function of the pi-terms from the scaling law, dependent on the fluid in the pressure sphere.

In Esparza and Baker (1977b), some tests were reported for glass spheres filled with Freon 12® at high pressure, but at the saturated <u>liquid</u> state rather than saturated <u>vapor</u>. Signals like damped sine waves were generated, and propagated radially without change in character, but with decrease in amplitude. The maximum pressure recorded, very close to a bursting sphere, was only 5.9 kPa (0.86 psig).

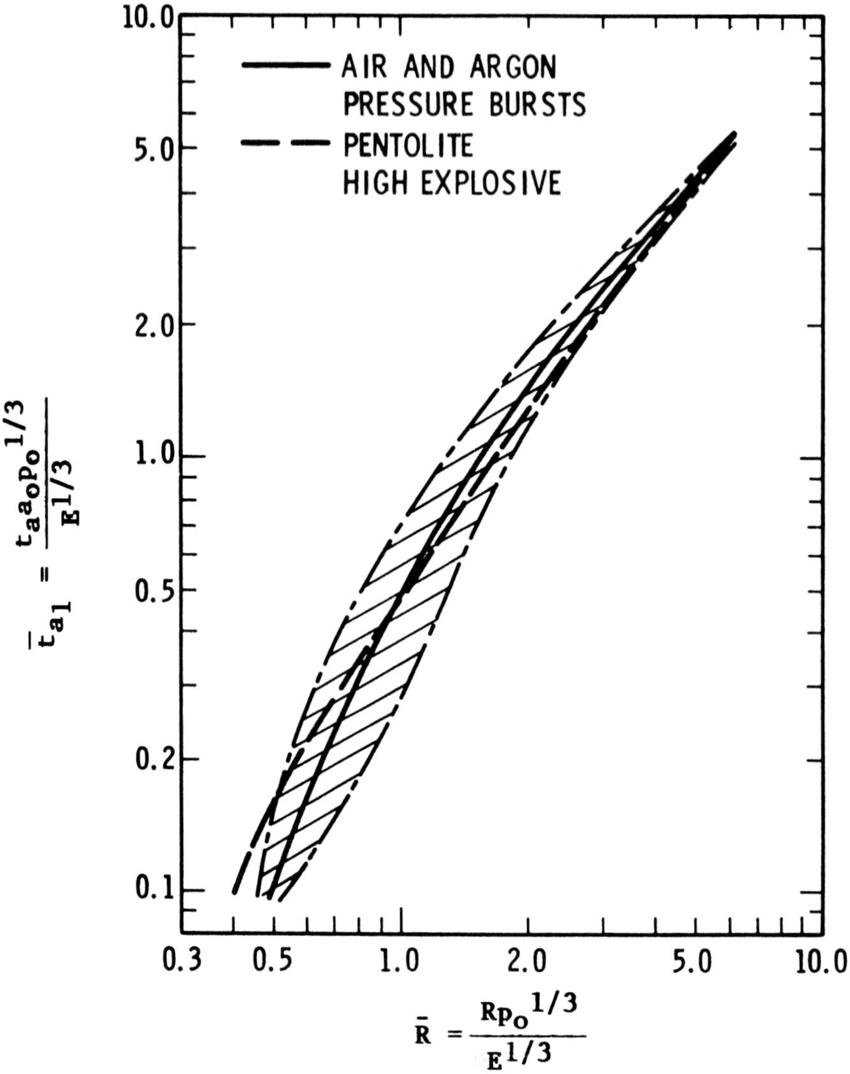

Fig. 5-18. Scaled time of arrival of first shock wave from bursting gas spheres. [Esparza and Baker (1977a)]

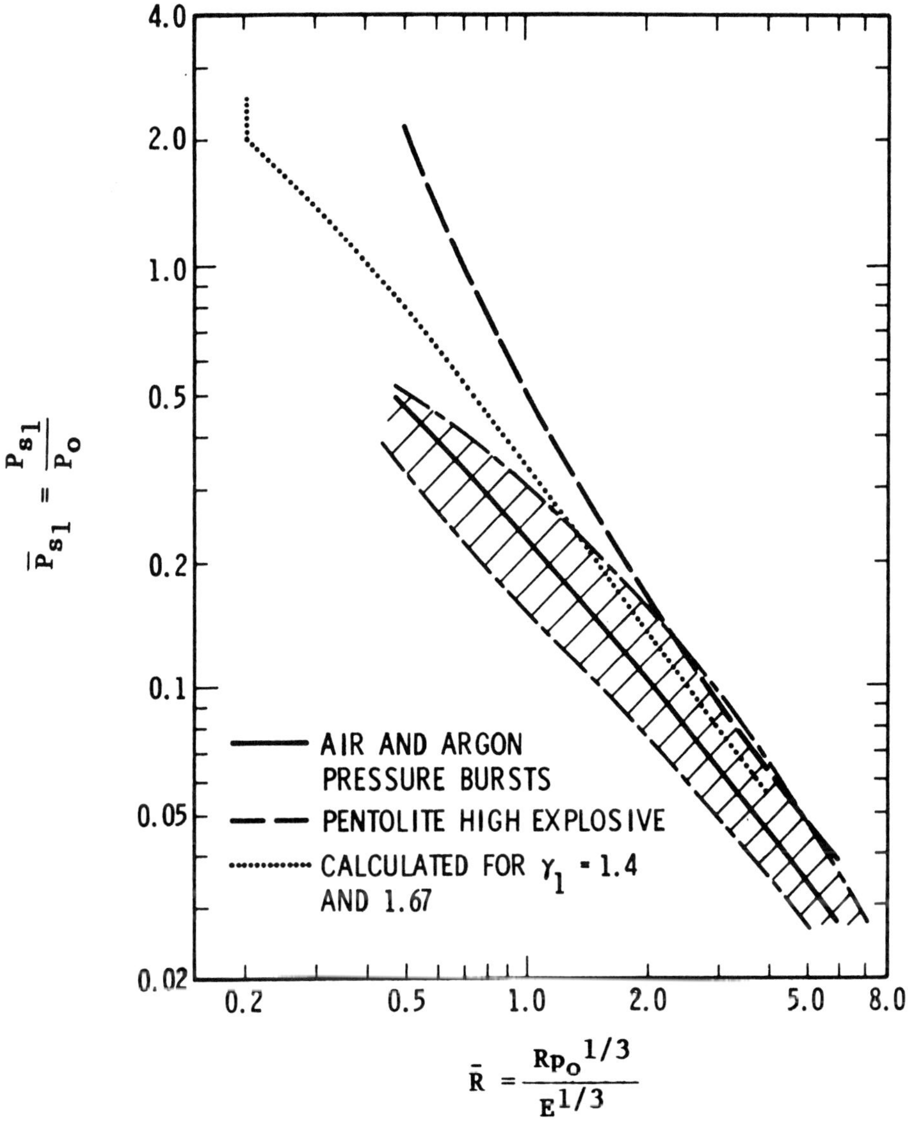

Fig. 5-19. Scaled side-on peak overpressure for first shock from bursting gas spheres. [Esparza and Baker (1977a)]

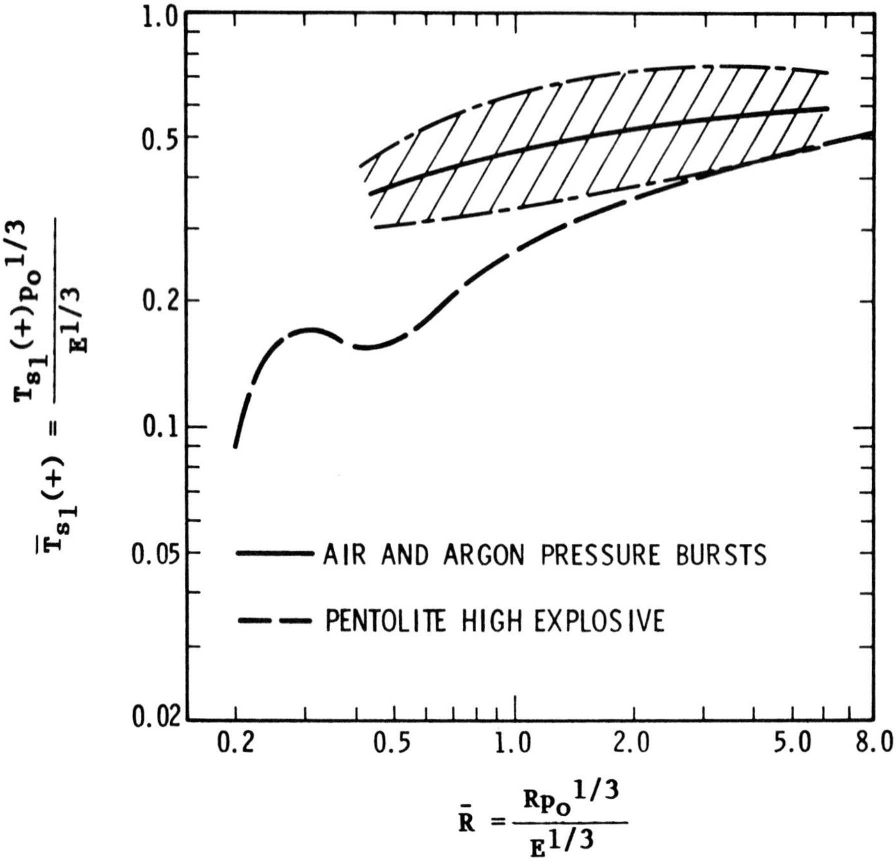

Fig. 5-20. Scaled duration of first positive phase of blast wave from bursting gas spheres. [Esparza and Baker (1977a)]

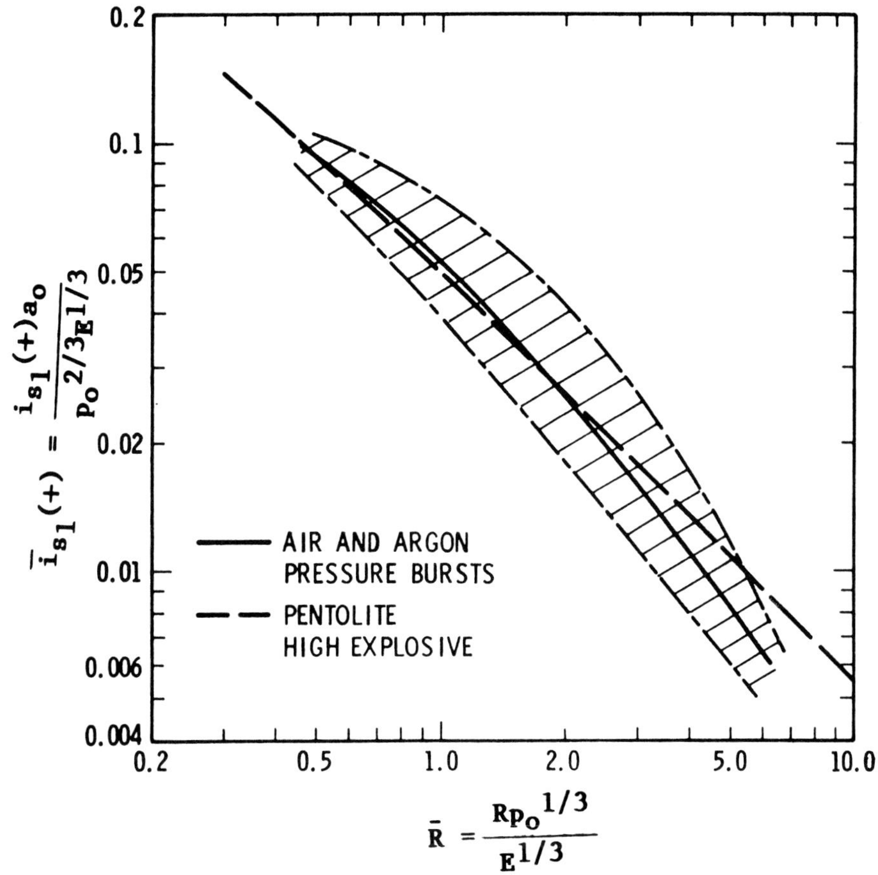

Fig. 5-21. Scaled side-on Positive impulse from bursting gas spheres. [Esparza and Baker (1977a)]

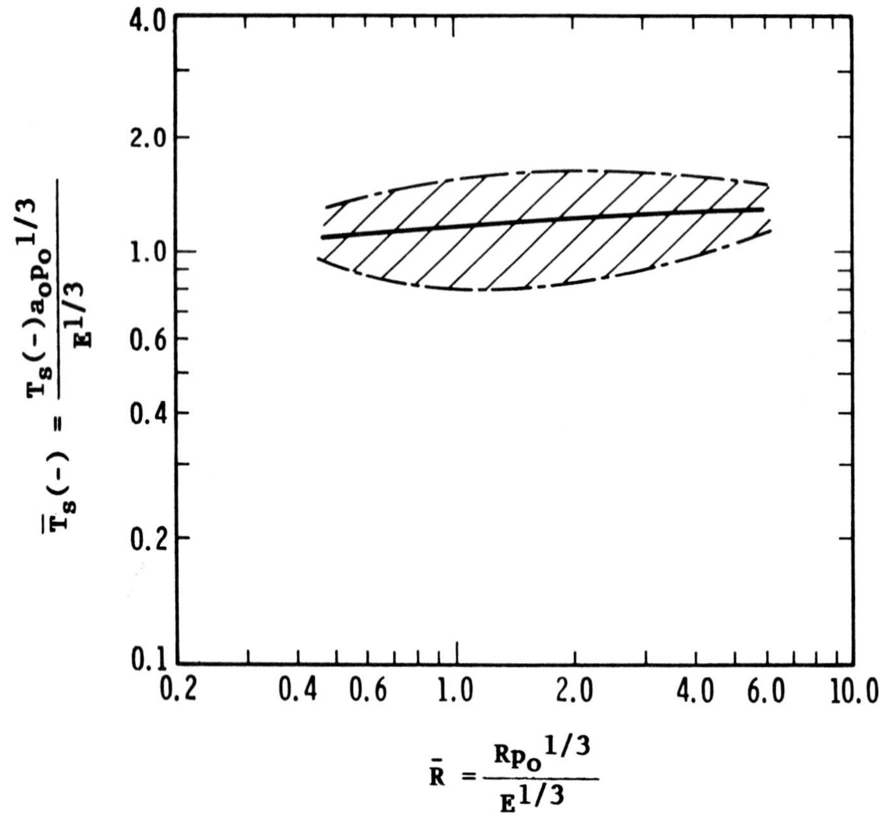

Fig. 5-22. Scaled duration of negative phase of blast wave from bursting gas sphere. [Esparza and Baker (1977a)]

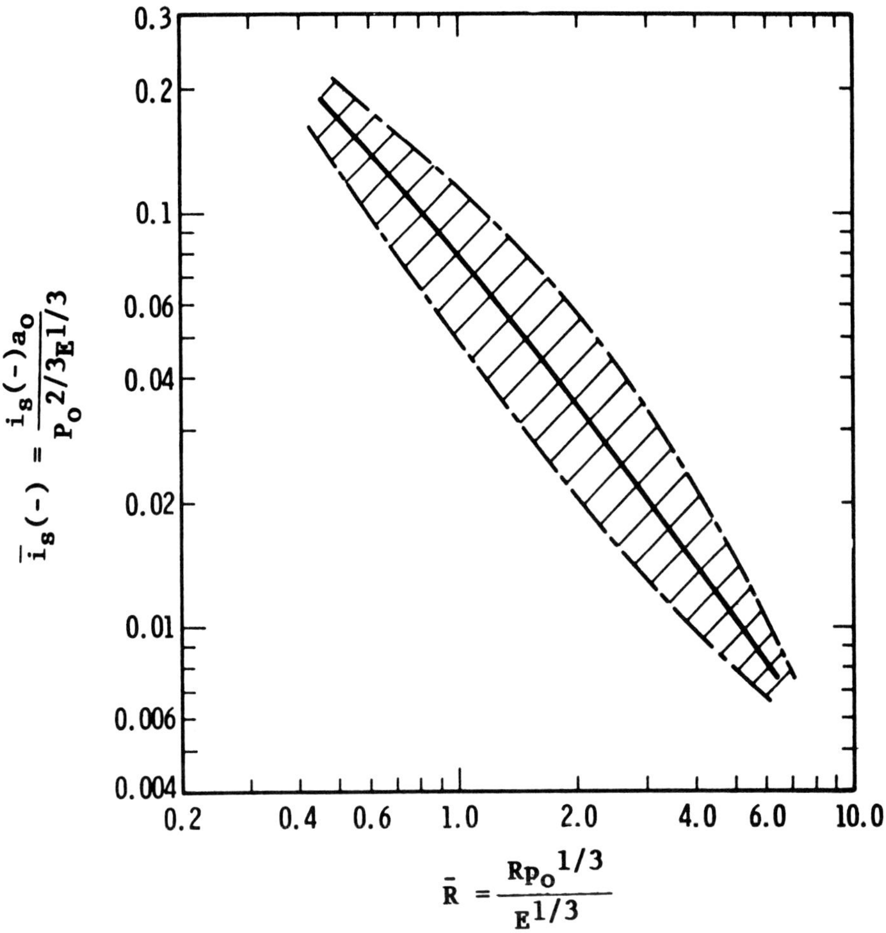

Fig. 5-23. Scaled side-on negative impulse from bursting gas spheres. [Esparza and Baker (1977a)]

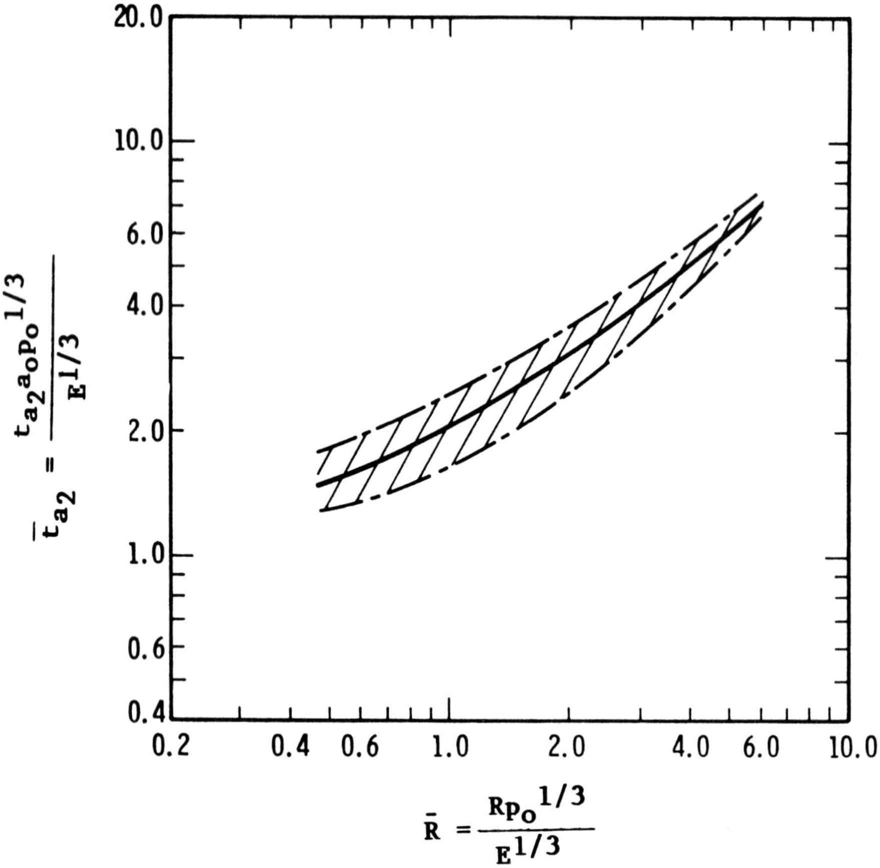

Fig. 5-24. Scaled time of arrival of second shock wave from bursting gas spheres. [Esparza and Baker (1977a)]

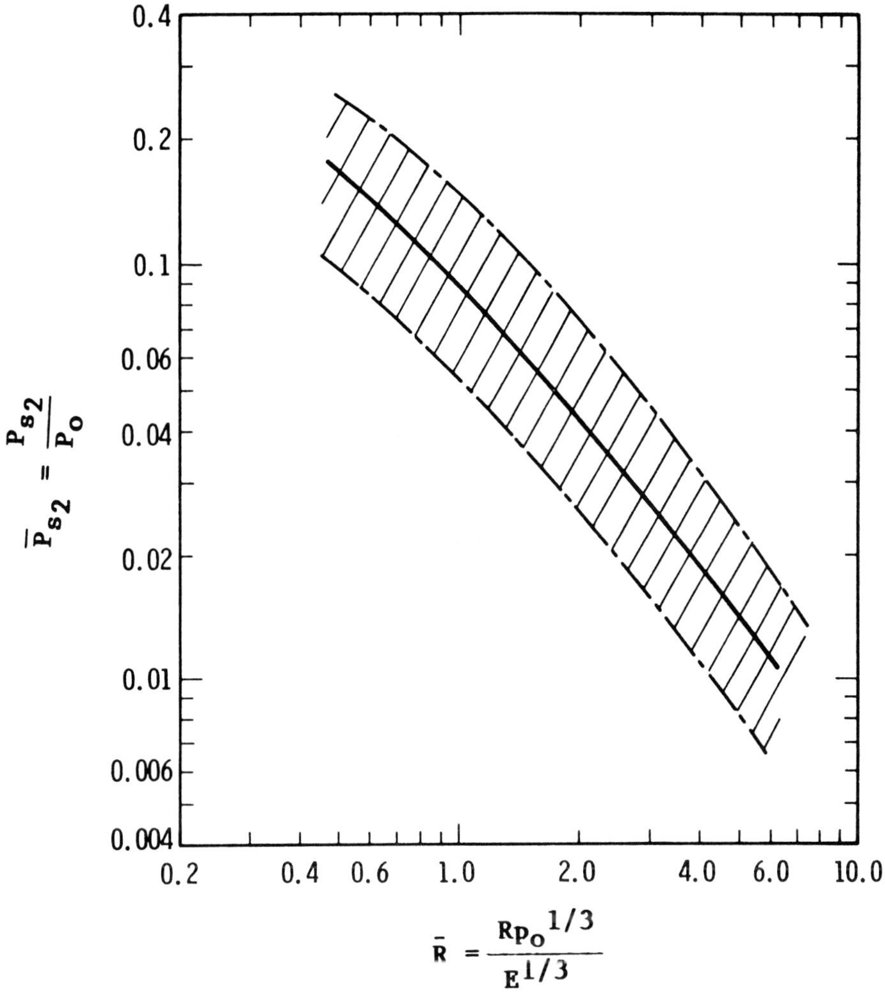

Fig. 5-25. Scaled side-on peak overpressure of second shock for bursting gas spheres. [Esparza and Baker (1977a)]

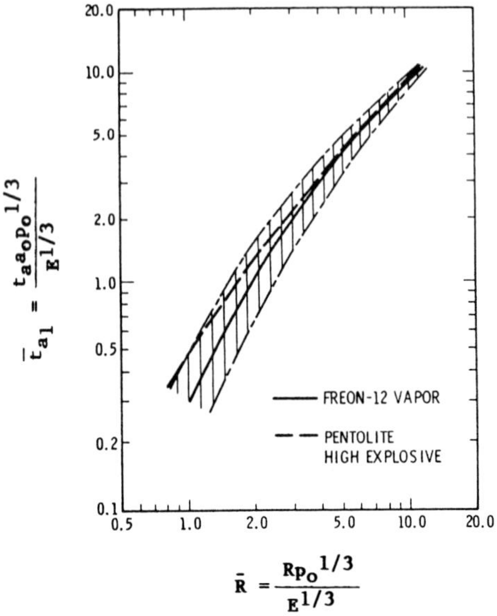

Fig. 5-26. Scaled time of arrival of first shock wave from bursting Freon-12® vapor spheres. [Esparza and Baker (1977b)]

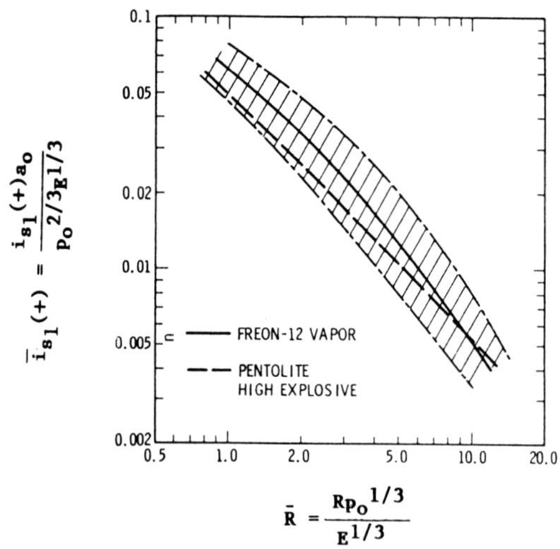

Fig. 5-27. Scaled side-on positive impulse from bursting Freon-12® vapor spheres. [Esparza and Baker (1977b)]

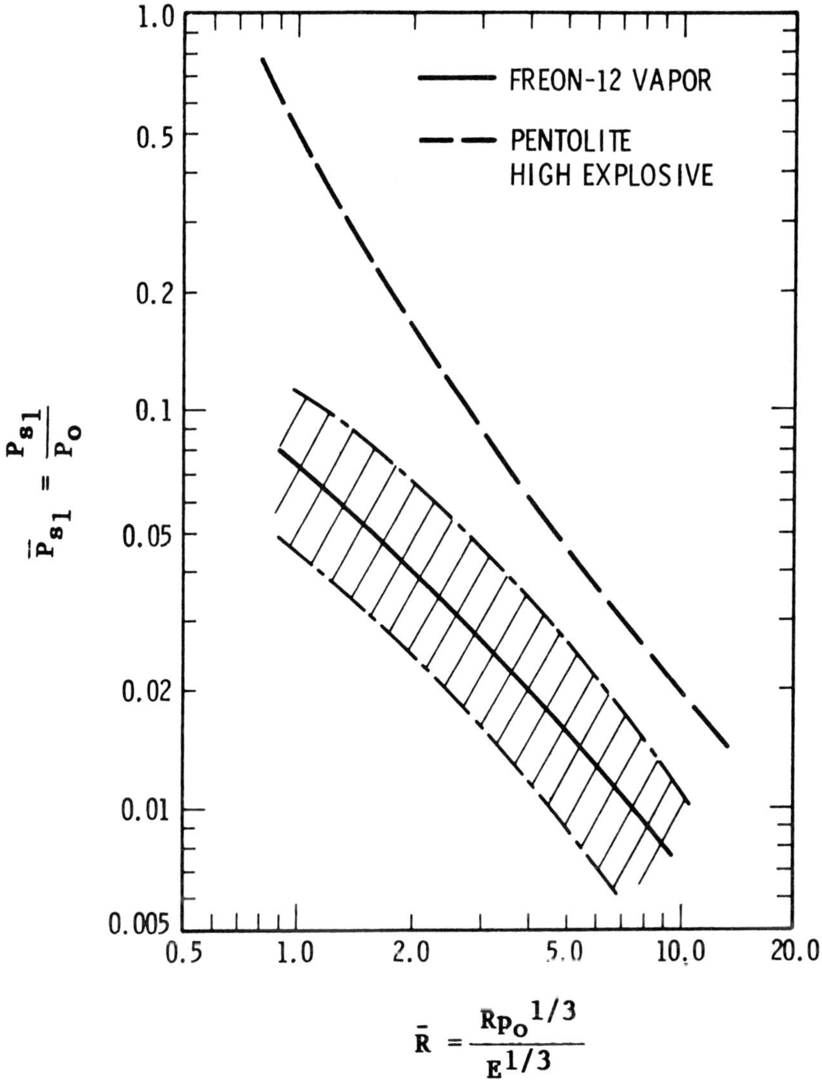

Fig. 5-28. Scaled side-on peak overpressure for bursting Freon-12® vapor spheres. [Esparza and Baker (1977b)]

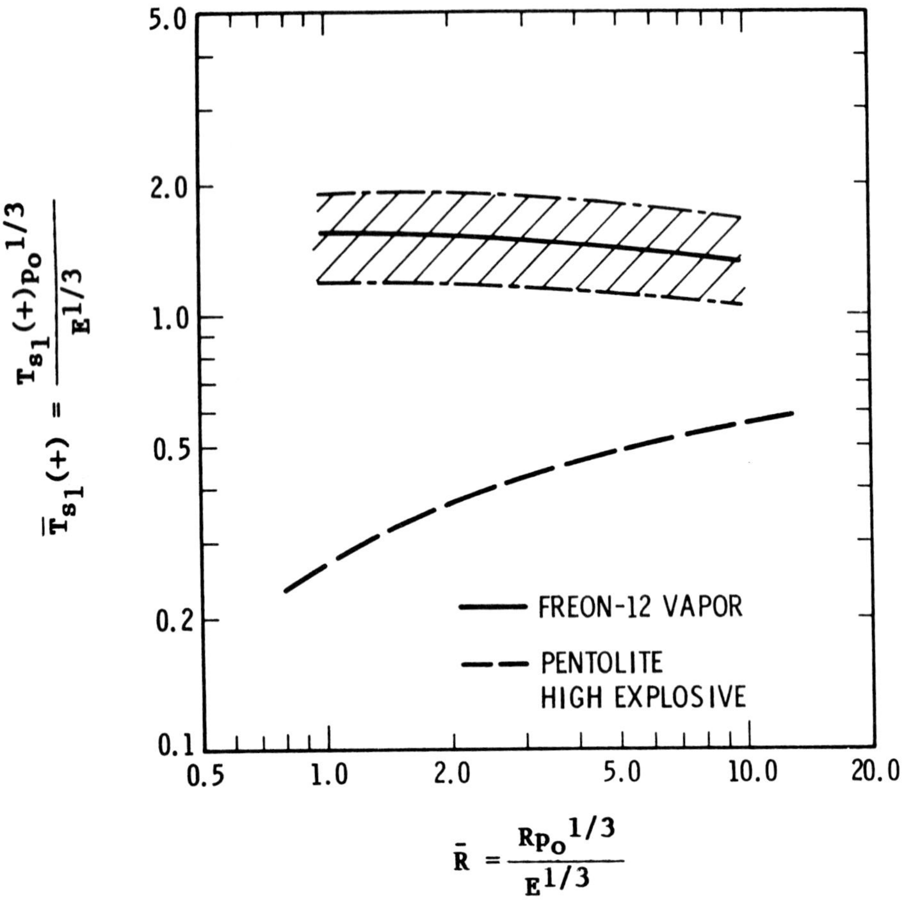

Fig. 5-29. Scaled duration of positive phase of blast wave from bursting Freon-12® vapor spheres. [Esparza and Baker 1977b)]

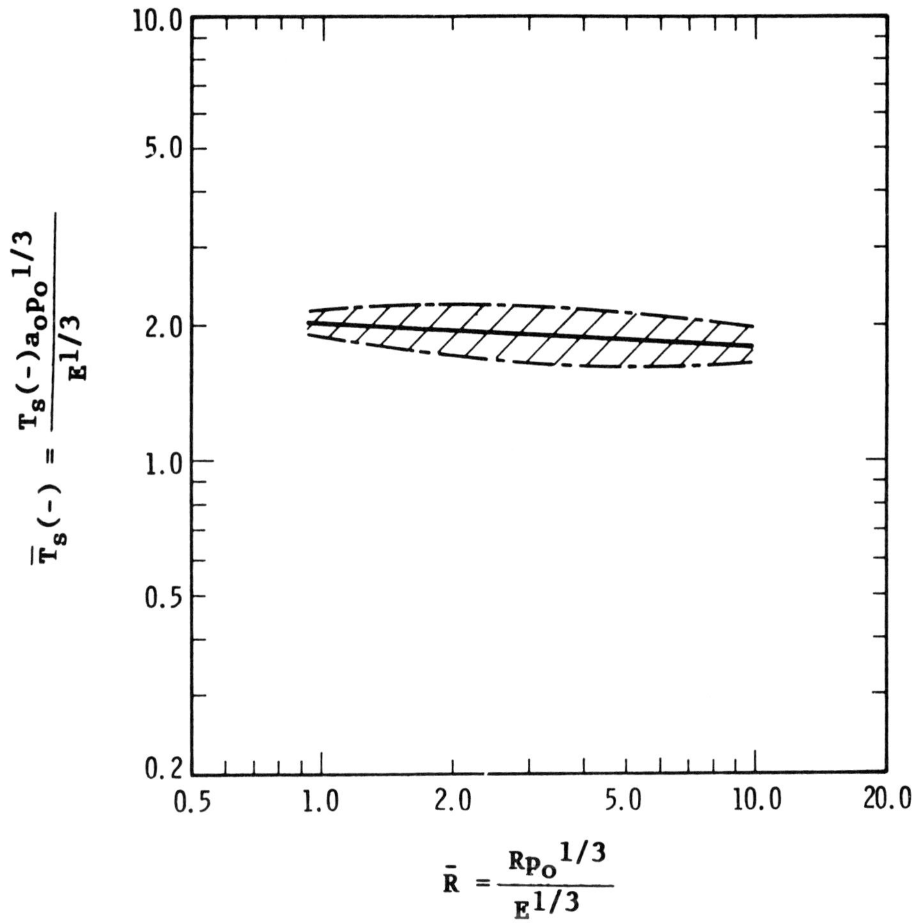

Fig. 5-30. Scaled duration of negative phase of blast wave from bursting Freon-12® vapor spheres. [Esparza and Baker (1977a)]

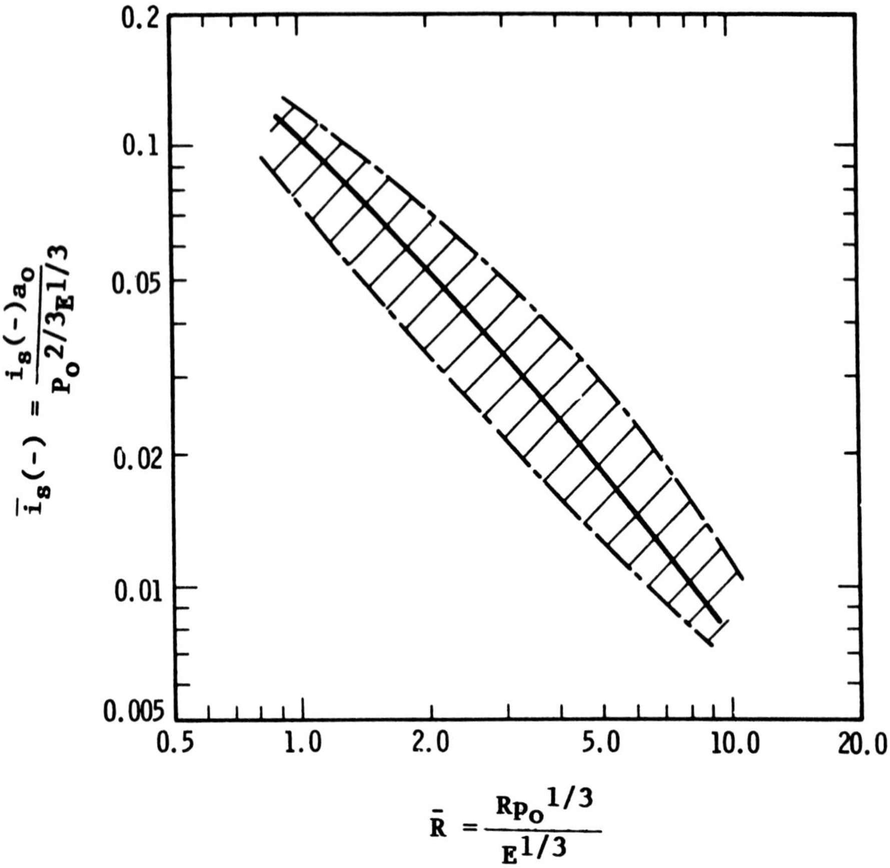

Fig. 5-31. Scaled side-on negative impulse from bursting Freon-12® vapor spheres. [Esparza and Baker (1977b)]

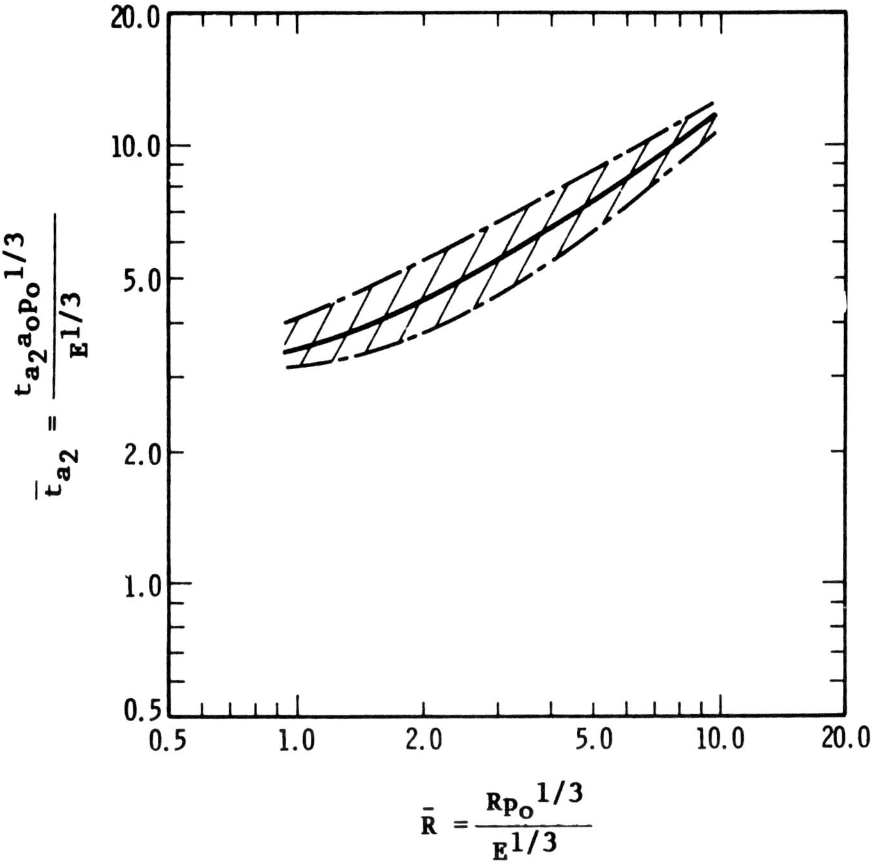

Fig. 5-32. Scaled time of arrival of second shock wave from bursting Freon-12® vapor spheres. [Esparza and Baker (1977b)]

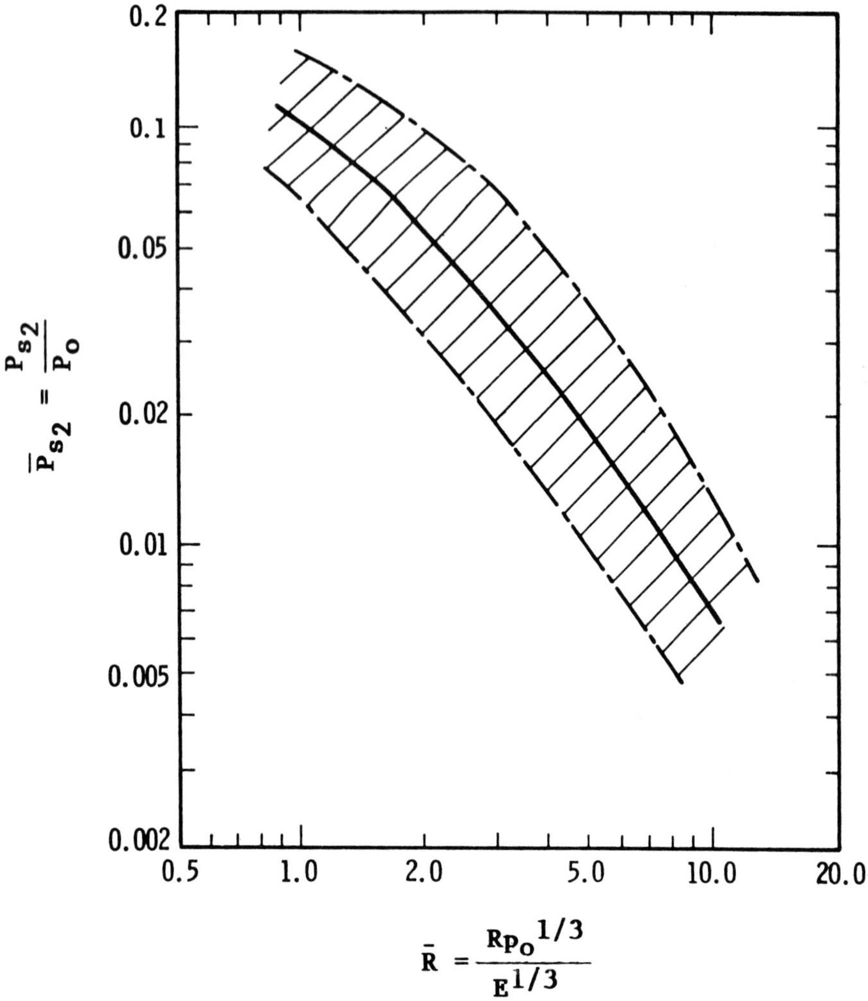

Fig. 5-33. Scaled side-on peak overpressure of second shock wave from bursting Freon-12® vapor spheres. [Esparza and Baker (1977b)]

In one other limited test program, some blast data were collected for bursts of small pressure cylinders partially filled with water and heated until they burst. [Baker, et al, (1978)]. The vessels were weakened so that they split longitudinally. The vessels were surrounded by a very light, expendable furnace and heated with natural gas flame. When they burst, they split longitudinally, and remained in one piece. Internal pressures and temperatures were recorded during heating, and blast pressures were sensed at burst by an array of side-on blast transducers. The resulting blast waves were somewhat similar to those from the glass sphere gas bursts, showing strong first shocks, pronounced negative phases, and second shocks. But, the second shocks were weaker than for the gas bursts. A typical pressure record is shown in Fig. 5-34. These explosions are, of course true steam explosions, as vessel explosions with all flash-evaporating fluids are often called. Only five burst tests were run during this project, but the authors did compare first positive phase blast parameters to curves from compiled TNT data, using the scaling law of Eq. (1) and the method noted earlier for calculating explosion source data, Figs. 5-35 and 5-36. Note the rather wide data scatter, but general agreement with the TNT curves over the range of the data.

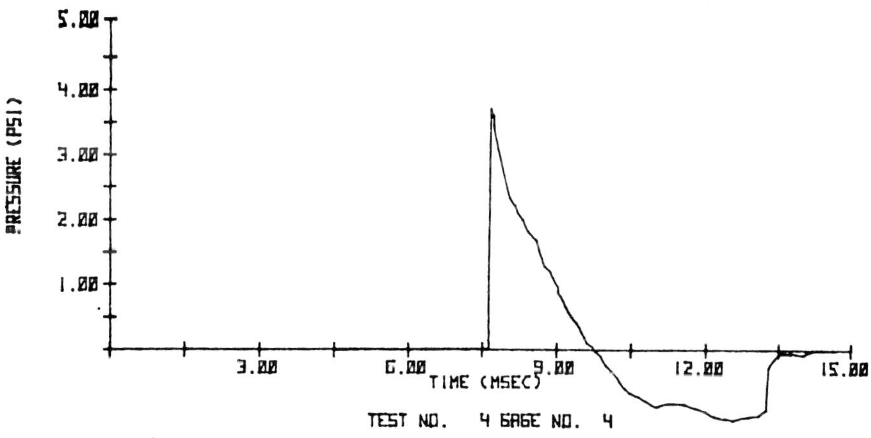

Fig. 5-34. Steam explosion blast record. [Baker, et al (1978)]

There has been little analysis or testing to better characterize blasts from bursting pressure vessels since about 1978. But recently, NASA and the U.S. Air Force have renewed their interest in better assessment of blast and fragment

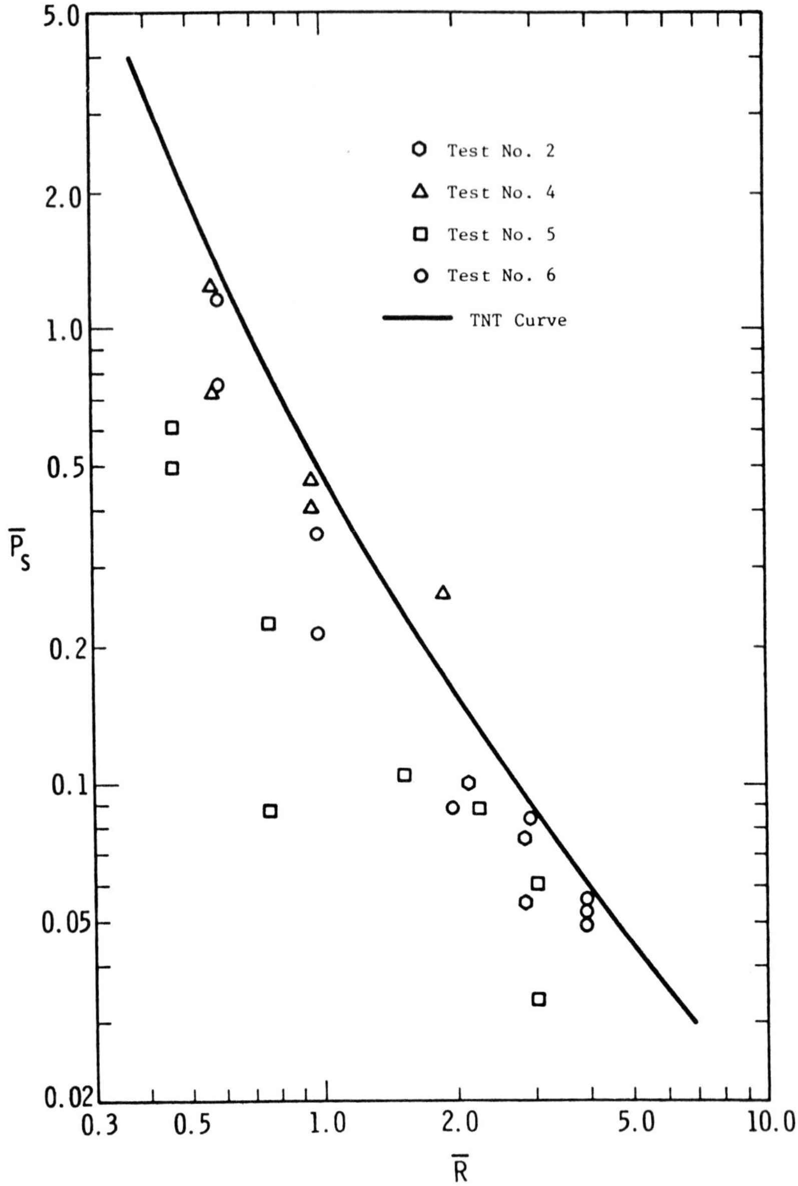

Fig. 5-35. Scaled side-on peak overpressures from bursting steam cylinders. [Baker, et al (1978)]

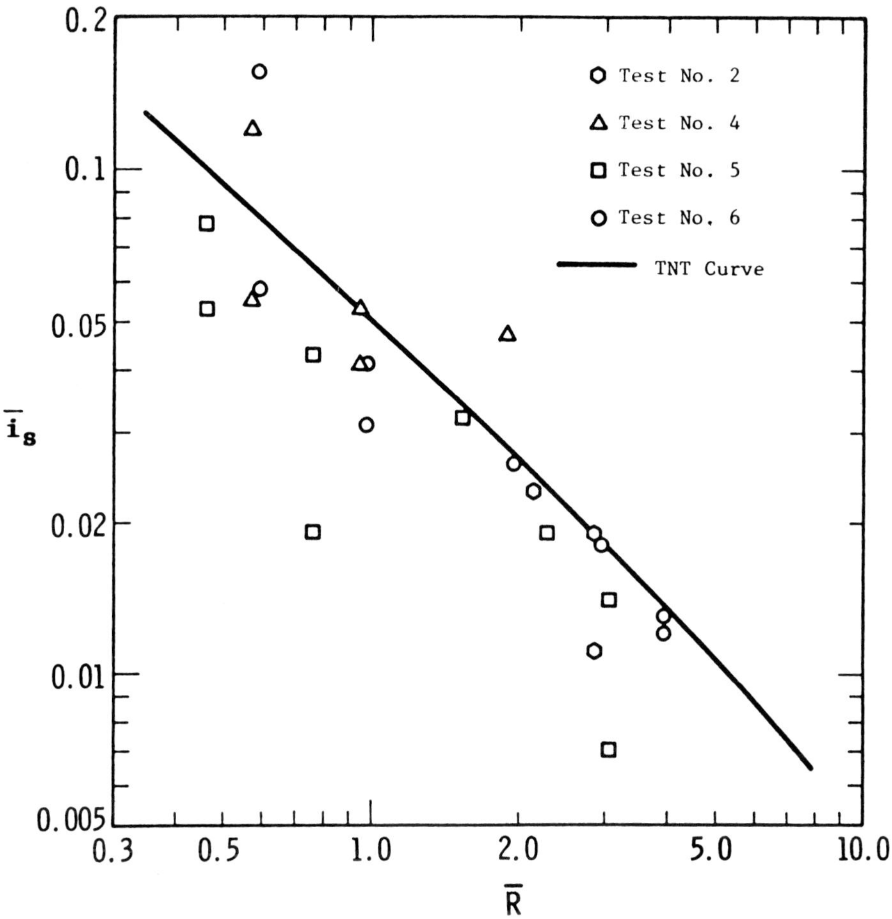

Fig. 5-36. Scaled side-on positive impulse from bursting steam cylinders. [Baker, et al (1978)]

effects for pressure vessel bursts. Coleman, et al (1988) present a rather thorough review of the past work on this topic, and also outline a test program designed to better characterize blast and fragments for spherical and cylindrical gas pressure vessels failed at pressures from 6.9 MPa to 69 MPa.

At high pressures, real gas effects alter behavior from the perfect gas equation of state. Intermolecular forces and finite molecular size cause these deviations. Wiedermann (1986) considers how these real gas effects alter explosion source energy for vessel bursts with high pressure gases. Fig. 5-37 illustrates this deviation for nitrogen at room temperature. Note that gas density becomes limited and less than that predicted by the perfect gas assumption. Sound velocity also increases significantly.

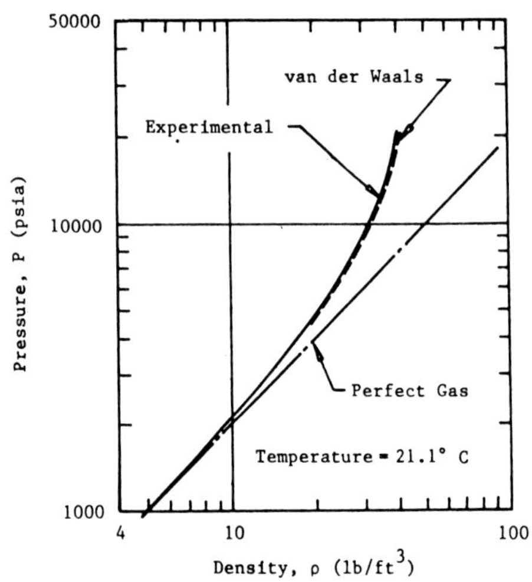

Fig. 5-37. Isotherm for nitrogen at high pressures. [Wiedermann (1986)]

Wiedermann (1986) gives first the equation developed by Baker (1983) for source energy based on isentropic expansion of a perfect gas,

$$E_1 = \frac{p_1 V_1}{(\gamma - 1)} \left[1 - \left(\frac{p_o}{p_1} \right)^{(\gamma-1)/\gamma} \right] \tag{14}$$

where p_1, V_1 are initial absolute pressure and vessel volume, p_o is ambient pressure and γ is ratio of specific heats for gas in the vessel. (this equation gives a slightly different but more accurate value for E than Eq. (3) which is based on total internal energy in the compressed perfect gas.) Then, he

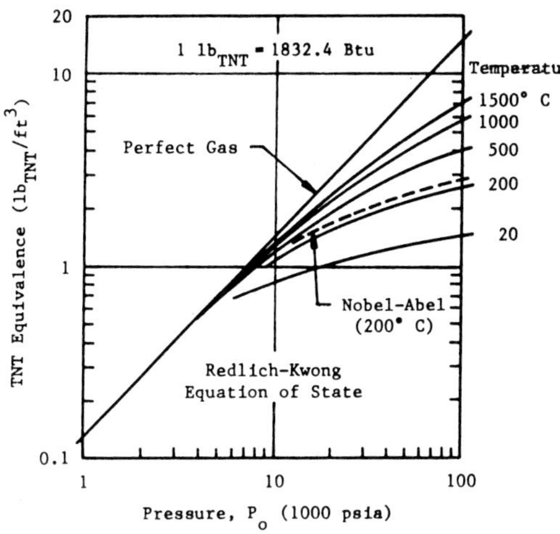

Fig. 5-38. TNT energy equivalence for pressurized vessels filled with argon. [Wiedermann (1986)]

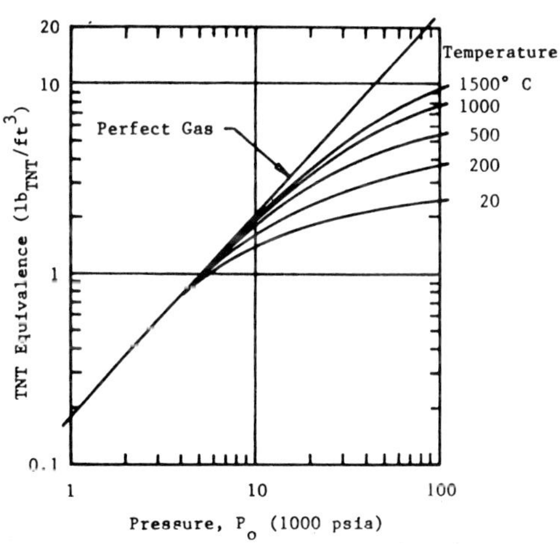

Fig. 5-39. TNT equivalence for pressurized vessels filled with nitrogen. [Wiedermann (1986)]

demonstrates that the isentropic expansion energy for real gases can be given by

$$E = (1 - \beta)\, E_1 \tag{15}$$

where

$$\beta = bp_o \tag{16}$$

and b is a covolume parameter measured experimentally or fitted to an equation of state. The parameter b has been measured or predicted for most gases.

Using these equations and literature data, Wiedermann predicts blast source energy per unit volume for argon and nitrogen versus pressure and temperature. Results appear in Figs. 5-38 and 5-39. Wiedermann apparently assumes that one would then use standard TNT blast curves to predict blast wave properties.

This recent work can be valuable if one must assess blast effects for bursts of very high pressure, gas vessels. Knowing initial conditions and vessel volumes, one can use curves such as Figs. 5-38 and 5-39 to calculate explosion source energy. For close-in effects, TNT equivalency is not appropriate, however, and curves such as Figs. 5-3 and 5-4 should be used to predict blast wave properties, rather than TNT blast curves as Wiedermann implies.

E. Closure

The author feels that he has covered most of the literature on analyses and testing of bursting pressure vessel explosions. Also included are numerous graphs of dimensionless blast wave properties. There has indeed been rather extensive effort to characterize this class of explosion, probably second only to analysis and testing of blast from high explosives. But, please realize that the effort is a very distant second. Whereas blast parameters for high explosives are based on countless thousands of well-instrumented tests, the data base for blasts from vessel bursts relies on only a few hundreds of tests. Only two sets of tests yield any detailed data on blast parameters past the first positive phase, even though the extant measurements show that distinctive negative phase and second shock properties result from vessel explosions.

Also, for those interested in determining how these blasts affect nearby structures or objects, there have been no measurements or analyses on interaction of these relatively complex waves with such targets.

It seems quite apparent that well-conducted tests to verify hydrocode predictions of vessel burst blast, as outlined by Coleman, et al (1988) are badly needed.

VI. LIQUID PROPELLANT EXPLOSIONS

A. Fundamentals

It may seem inappropriate to include the topic of liquid propellant explosions in this book, because such explosions are not truly gas or hybrid explosions. But, for reasons we will enumerate here, accidental explosions involving the common liquid rocket propellants are quite variable and the explosion sources are often very heterogeneous. So, this "ill-conditioned" class of explosions has much more in common with industrial gas or hybrid explosions than the much better defined high explosive detonations.

Because huge quantities of the energetic materials which constitute the rocket propellants are required to launch large liquid fueled rockets, there is great potential for serious explosions in the event of certain system failures on or near the launch pad. Almost all liquid rocket propellants consist of two liquid components, a fuel and an oxidizer. In normal operation, these two components are injected into a combustion chamber in the rocket in approximate stoichiometric ratio, burned there to increase gas pressure and temperature, and then exhausted through a nozzle or nozzles to produce thrust.

There are only a very limited number of liquid fuel-oxidizer combinations in practical use in rocketry. The three most common are:

1) Liquid hydrogen (LH_2) and liquid oxygen (LO_2)
2) Kerosene (RP-1) and liquid oxygen (LO_2)
3) 50% unsymmetrical dimethyl hydrazine (UDMH) - 50% hydrazine (N_2H_4) and nitrogen tetroxide (N_2O_4)

For an explosion to occur, there must usually be some failure of or leakage from <u>both</u> the fuel and oxidizer tankage, violent mixing of these materials, and a delayed ignition. Severity of a resulting explosion (generating a damaging blast wave) can range from minimal to catastrophic. Liquid hydrogen and liquid oxygen are cryogenic materials, and can only be kept in their liquid states in well-insulated tankage. The remaining fuels and oxidizers are all liquids with low vapor pressures at normal ambient temperatures. The first two fuel-oxidizer combinations must be ignited by some external source after mixing to start an explosion (or simply combustion). The third combination is, however, <u>hypergolic</u> and so ignites spontaneously as the two materials reach a flammable ratio during mixing.

In the stoichiometric ratios just required for complete combustion, the mass ratios of fuel to oxidizer and the theoretical heat of combustion on a mass basis are given for the three most common liquid propellants combinations in Table 6-1.

For both propellant combinations involving cryogenic materials, intimate mixing in the manner of readily miscible liquids at the same initial temperatures is impossible. For the LH_2/LO_2 combination, only very heterogenous "mixtures" are possible, because LH_2 temperature is below the freezing point for LO_2, and conversely, LO_2 temperature is above the boiling point for LH_2. Furthermore, LO_2 is much denser than LH_2. So, when these two liquids are thrown together, LH_2 boils and LO_2 freezes. In general, the "mixture" will contain gaseous and liquid hydrogen, and liquid and solid oxygen. If this process occurs in air, the gaseous hydrogen may also form a flammable mixture with oxygen in the air. The situation is not quite so violent if LO_2 and RP-1 are thrown together, but the LO_2 will boil and the RP-1 will freeze. So, again no homogeneous mixture is possible.

TABLE 6-1. OPTIMUM FUEL-OXIDIZER COMBINATIONS FOR LIQUID PROPELLANTS

Fuel	Oxidizer	Mass Ratio, Fuel/Oxidizer	Heat of Combustion, H_c, kJ/kg
LH_2	LO_2	1/5	16,700
RP-1	LO_2	1/2.25	5,650
50% N_2H_4 - 50% UDMH	N_2O_4	1/2	10,800

Only in very small-scale experiments, where rapid mixing of LH_2 and LO_2 has been promoted by explosive means [A.D. Little (1960)], have energy releases per unit mass approaching that in Table 6-1 been achieved. Nothing approaching a condensed phase <u>detonation</u> has been observed, even in very small-scale experiments designed to trigger such a phenomenon [Tarifa and Perez del Notario (1966)].

Farber and his coworkers have shown that, for the first two propellant combinations in Table 6-1, there is most probably a natural internal ignition source of electrostatic discharge when cryogenic liquid propellants are violently mixed [Farber, et al (1968), (1965a), (1967a), (1967b), (1967c), (1965b), (1966a), (1966b), (1965c)]. So, given a large enough quantity mixing, ignition is almost certain (but the exact <u>time</u> and <u>location</u> of ignition can be quite variable.) Furthermore, for the hypergolic propellant combination, early ignition during mixing is assured. So, it is clear that it is quite inappropriate in estimating the characteristics of liquid propellant explosions to simply assume that all of the inventory of propellants can be mixed, and then exploded to realize the full explosion potential implied by theoretical maxima based on calculated heats of combustion.

These limitations were not fully realized when large liquid fueled rockets were being developed in the United States, and much of the launch facility siting at Kennedy Space Center was based on the worst-case assumption that all of the propellants in a launch-pad accident of a Saturn V rocket[1] could mix and detonate on the launch pad. But concurrently, the U.S. Air Force and NASA undertook an extensive test program with limited analytic support, under the name Project PYRO, to simulate and measure the blast and thermal effects of launch pad accidents involving the three propellant combinations in Table 6-1. Results of this program were reported in a three-volume final report [Willoughby, et al (1968)]. This excellent set of reports remains the best source for data on effects of liquid propellant explosions.

Later, NASA sponsored several small projects to cast the results of Project PYRO and other related work in the form of workbooks which could allow safety engineers to predict both blast and fragment effects from launch pad explosions [Baker, et al (1974), Baker, et al (1977), Baker, et al (1978)]. This chapter is largely a distillation of the NASA and Air Force-sponsored work, with a few new insights gathered in a recent study of explosion hazards for the Space Shuttle [NUS-4543 (1985)].

B. Classes of liquid propellant explosions

In the planning for Project PYRO, the overall concept of failures leading to explosions is shown in Fig. 6-1. The objectives of the explosive testing were first, to determine which parameters of potential interest in the interaction of two propellant masses had a significant influence on the explosive yield time relationship and, second, to determine the quantitative nature of the influence.

In trying to satisfy the above objectives, the PYRO group evolved the concepts of:

1) Testing at several scales;
2) Simulating interior bulkhead failures to promote mixing as one serious failure mode;
3) Simulating nearly simultaneous tankage rupture for cryogenic propellants on pad fall-back as a second serious failure mode; and
4) Simulating high-velocity impact (as from early guidance failure) as a third serious failure mode.

The three classes of explosion for the extensive test series was then designated:

1) Confined-By-Missile (CBM)
2) Confined-By-Ground-Surface (CBGS)
3) High-Velocity-Impact (HVI)

---

[1]The Saturn V contained about 2,700,000 kg of liquid propellants.

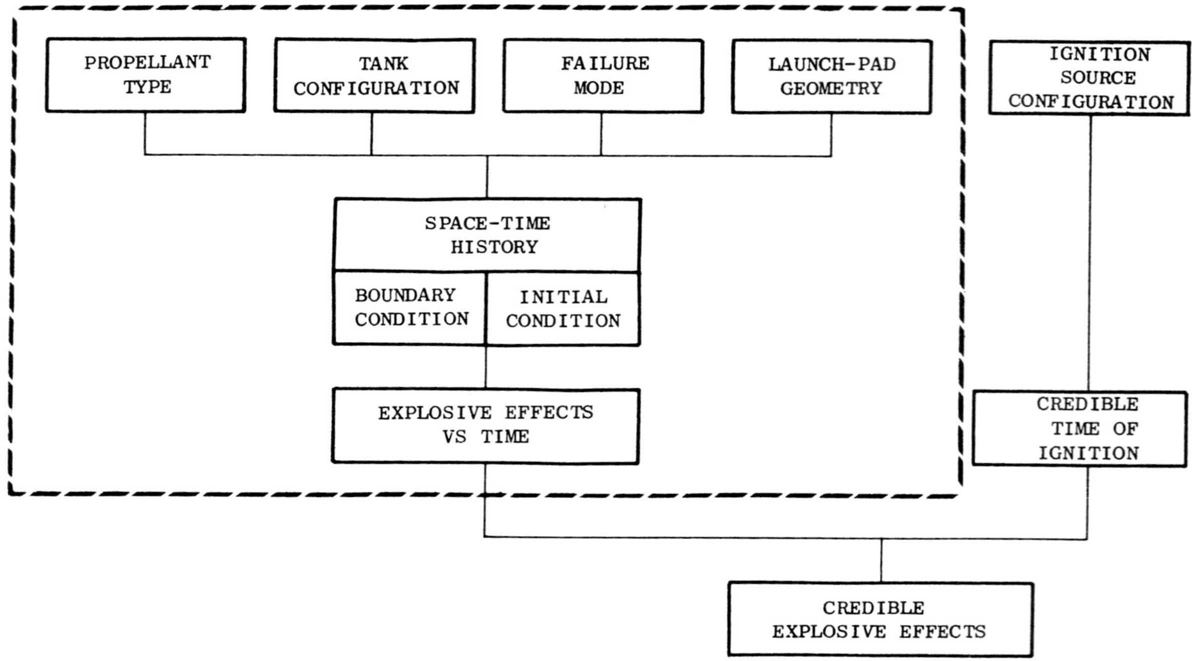

Fig. 6-1. Generalized concept of the missile failure and explosive process. [Willoughby, et al (1968)]

In the testing program, aluminum alloy tanks of various sizes were made and tested at a special test facility set up at Edwards Air Force Base, California, and at the K2 rocket track at Naval Weapons Center, China Lake, California. Primary instrumentation consisted of many channels of air blast gages, and extensive thermal response instrumentation. Multiple motion picture coverage was also provided at several framing rates, primarily for viewing fireball growth and durations, but also later used for obtaining data on fragmentation [Baker, et al (1974)].

For the cryogenic propellants, three total propellant weights were chosen for test; 90.7, 453 and 11,300 kg. The basic tank configuration for CBM tests is shown in Fig. 6-2. Two different L/D ratio tanks were tested; 5:1 and 1.8:1. The size of the circular opening sheared out by the explosively-driven striker was either the full tank diameter, or 0.45 of tank diameter. General tank configurations were similar for CBGS tests, except that tank closures were flat diaphragms, and a double diaphragm shear mechanism was used to rupture both tanks after guided free fall. Fig. 6-3 shows the tank arrangement and the six-bladed shears for low drop height CBGS tests,while Fig. 6-4 shows the special tower constructed at the Edwards AFB test site for drops from greater heights.

In high velocity impact (HVI) testing at China Lake, propellant tankage was also aluminum alloy, with general configuration similar to Fig. 6-5 for all propellant combinations. These tanks were mounted on slippers and accelerated by solid rocket motors on the K2 track to velocities between 100 m/s and 270 m/s. They were then allowed to fly free for a short distance before impacting on several types of solid targets which simulated several possible impact situations. As an example, one such target shown in Fig. 6-6 was intended to simulate explosions after impact on soft earth with partial penetration before explosion.

The primary instrumentation for all Project PYRO tests consisted of a blast pressure measuring system. Three gage lines were arranged for all tests, with somewhat different layouts for the CBM and CBGS tests at Edwards and the HVI tests at China Lake. Gage layouts are shown in Figs. 6-7 and 6-8. Several types of gage mounts were used, as shown in Figs. 6-9 through 6-11. The Type A mounts were employed at stations close to ground zero; Type B mounts on the concrete test pad at Edwards, and Type C mounts at greater distances. Transducers were all Kistler piezoelectric pressure transducers, with suitable amplification and recording on FM tape recorders to assure 70 kHz system frequency response. Not all gage locations were instrumented in every test, but locations were chosen to obtain measurements at about the same Hopkinson-Cranz scaled distances, as total propellant weights changed. Actual gage locations used in CBM and CBGS tests are summarized in Table 6-2.

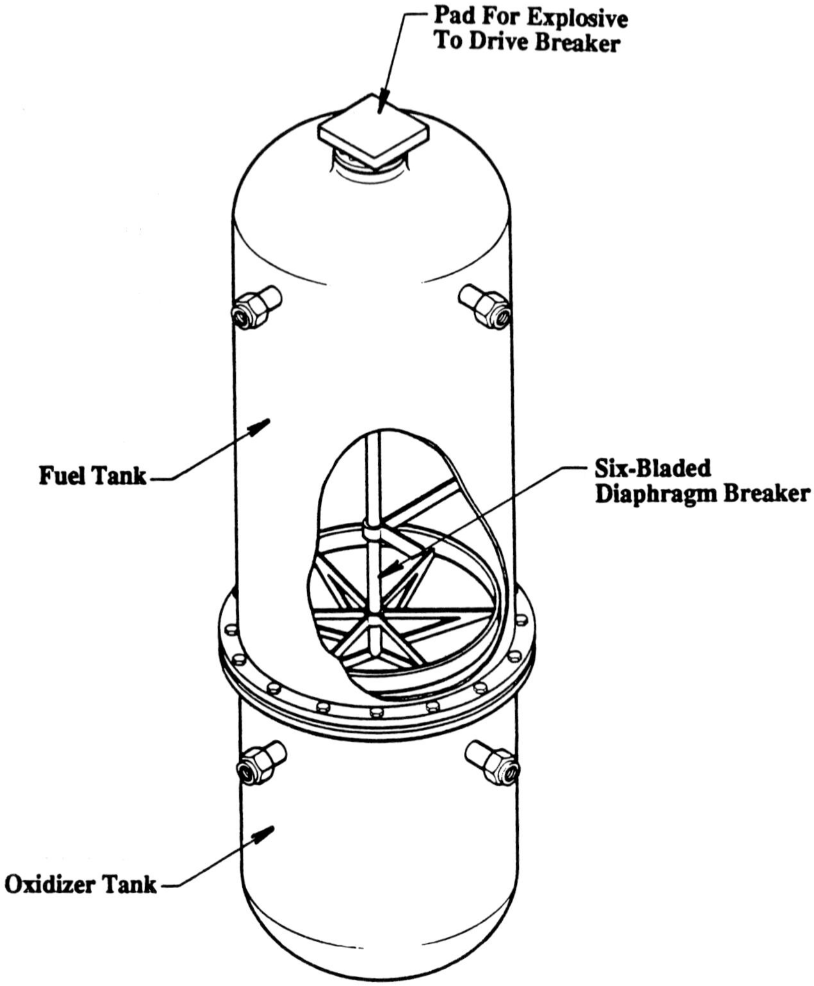

Fig. 6.2. Cryogenic confinement-by-the-missile (CBM) tank. [Willoughby, et al (1968)].

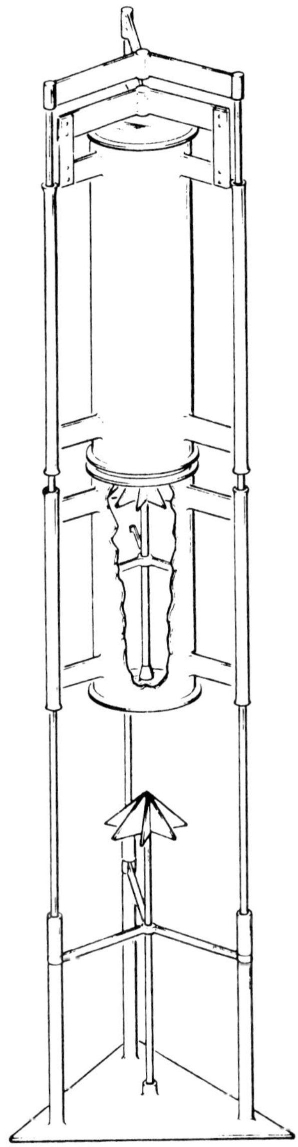

Fig. 6-3. Sketch of tank assembly and drop frame for the confinement-by-the-ground surface (CBGS) low drop test series. [Willoughby, et al (1968)]

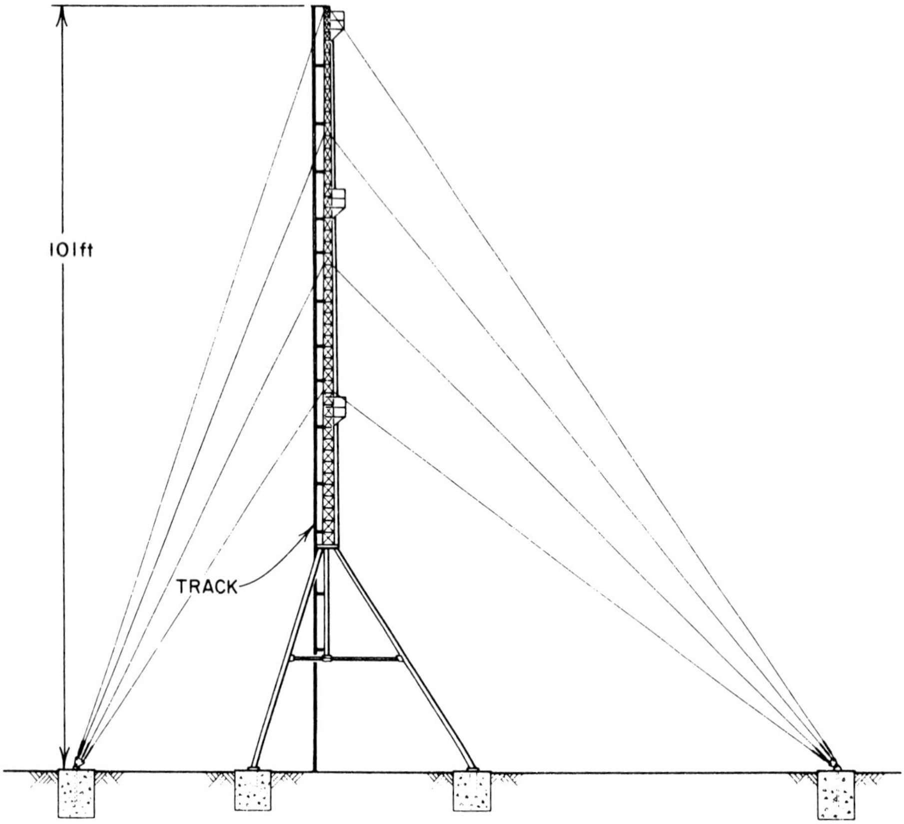

Fig. 6-4. Sketch of the drop tower. [Willoughby, et al (1968)]

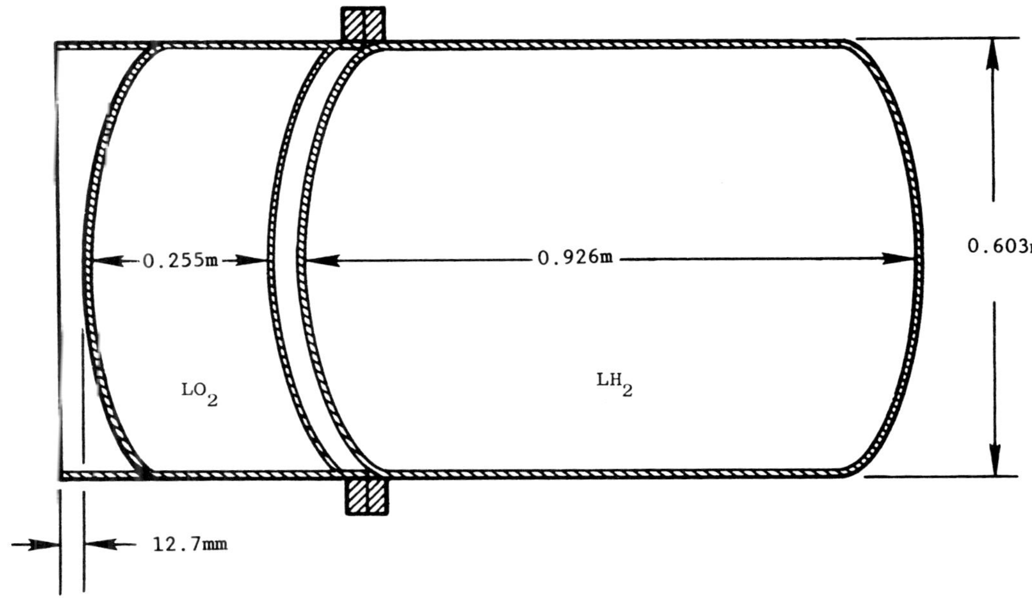

Fig. 6-5. 200-lb LO_2/LH_2 impact tank. [Willoughby, et al (1968)]

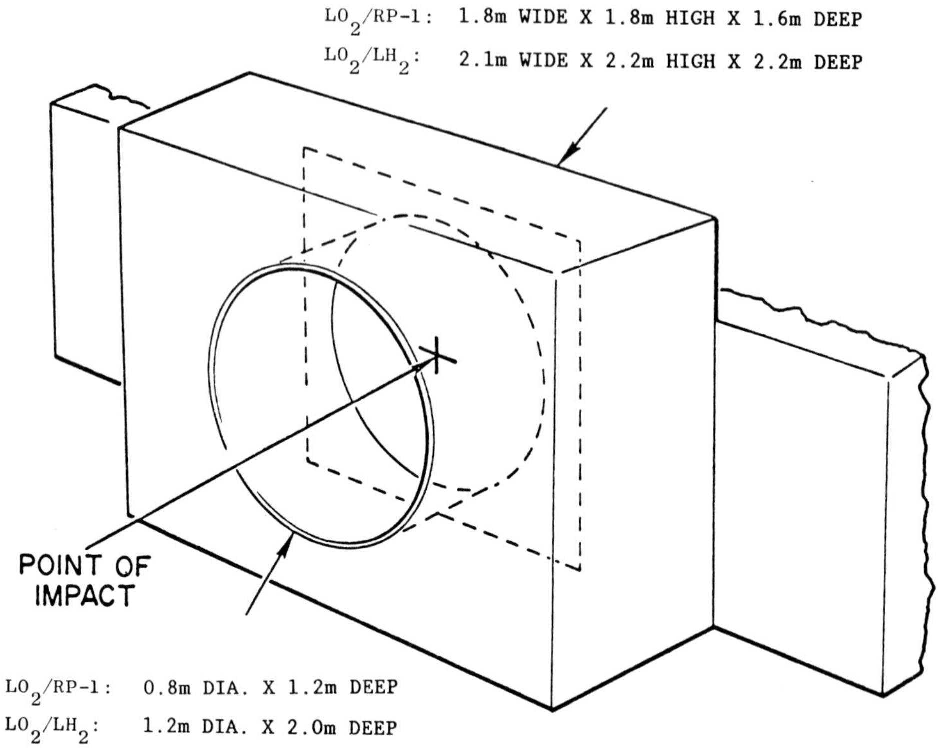

Fig. 6-6. Deep-hole target for the cryogenic impact tests. [Willoughby, et al (1968)]

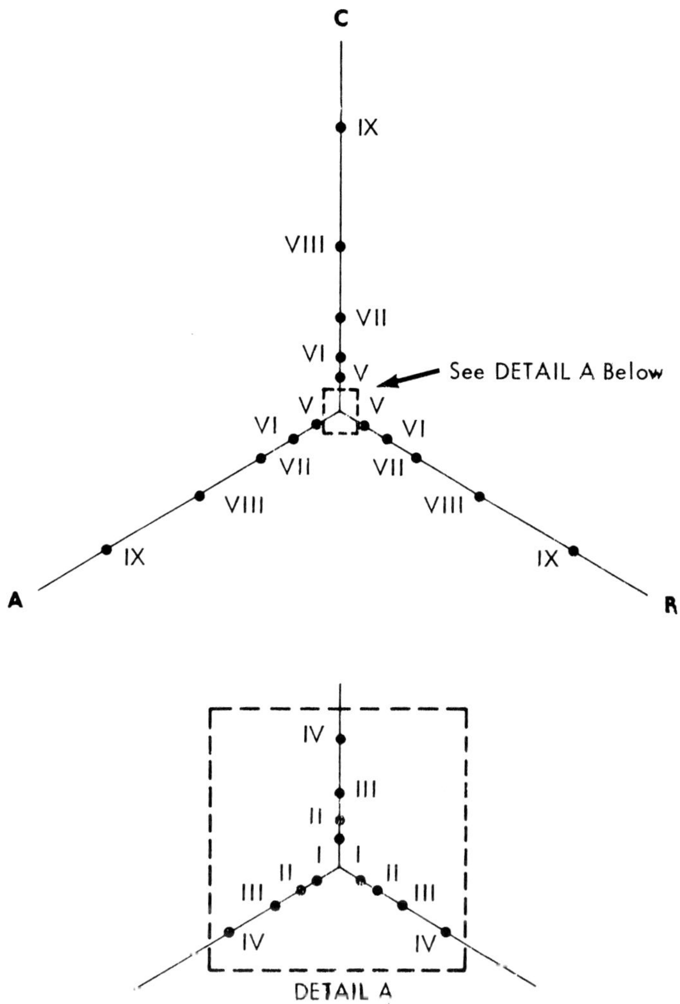

Fig. 6-7. Instrumentation layout for CBM and CBGS tests at Edwards AFB. [Willoughby, et al (1968)]

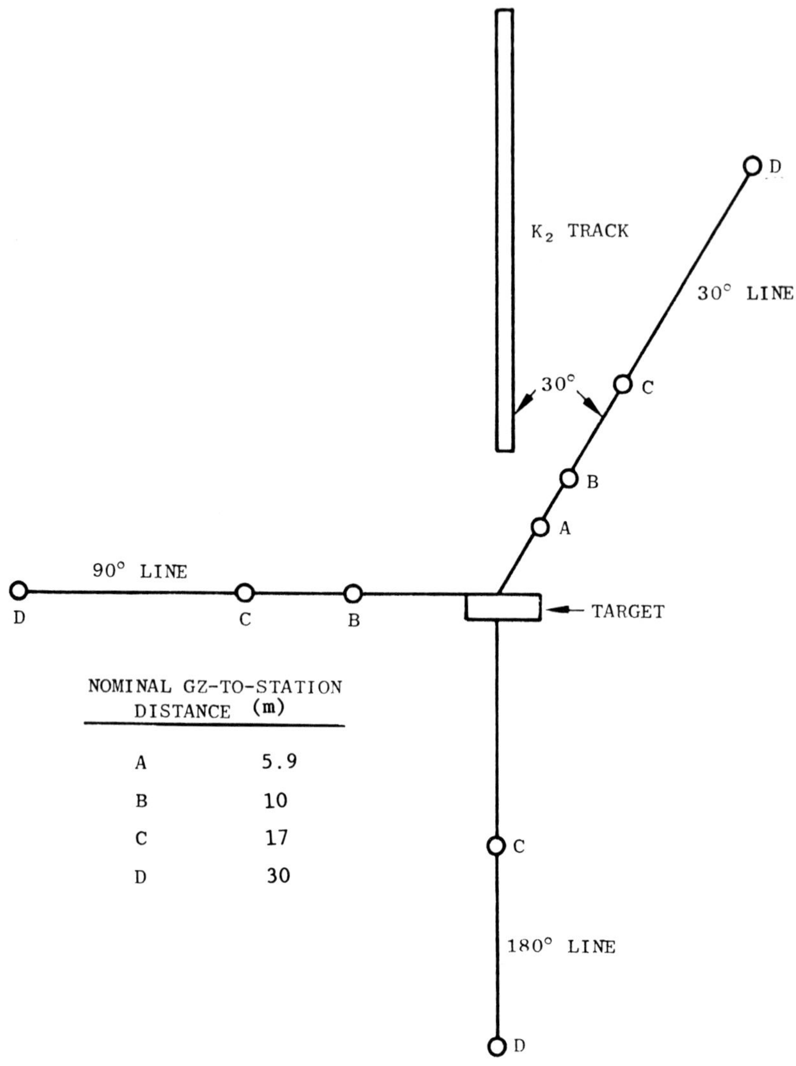

Fig. 6-8. Pressure sensor locations for majority of hypergolic impact tests at NWC K2 track. [Willoughby, et al (1968)]

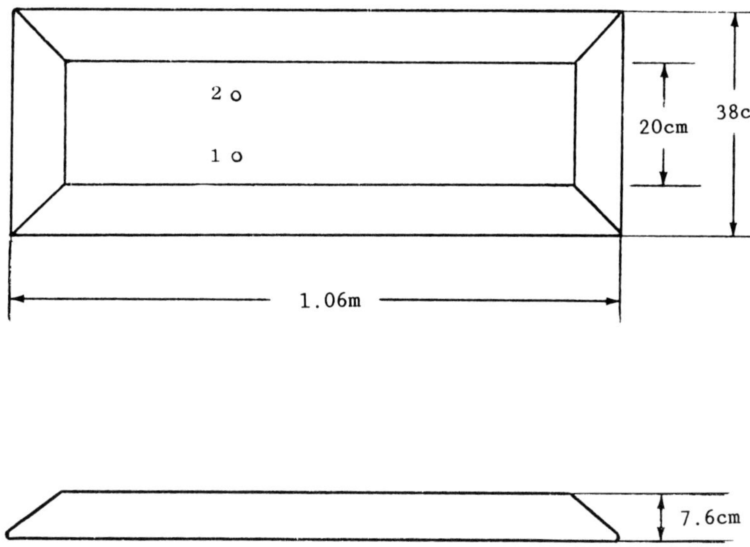

1. Side-on overpressure gauge

2. Stagnation gauge

Fig. 6-9. Type A sensor mount. [Willoughby, et al (1968)]

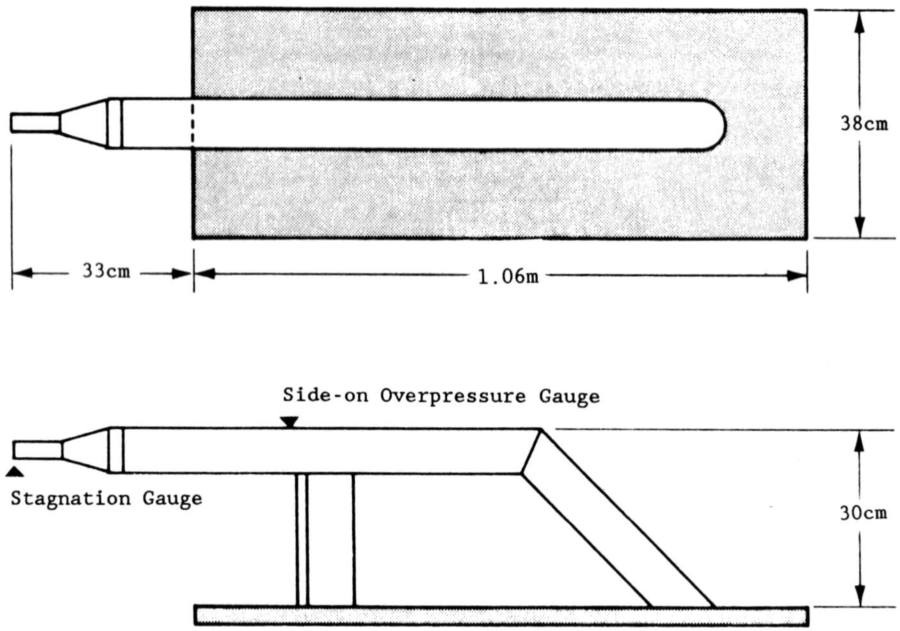

Fig. 6-10. Type B sensor mount. [Willoughby, et al (1968)]

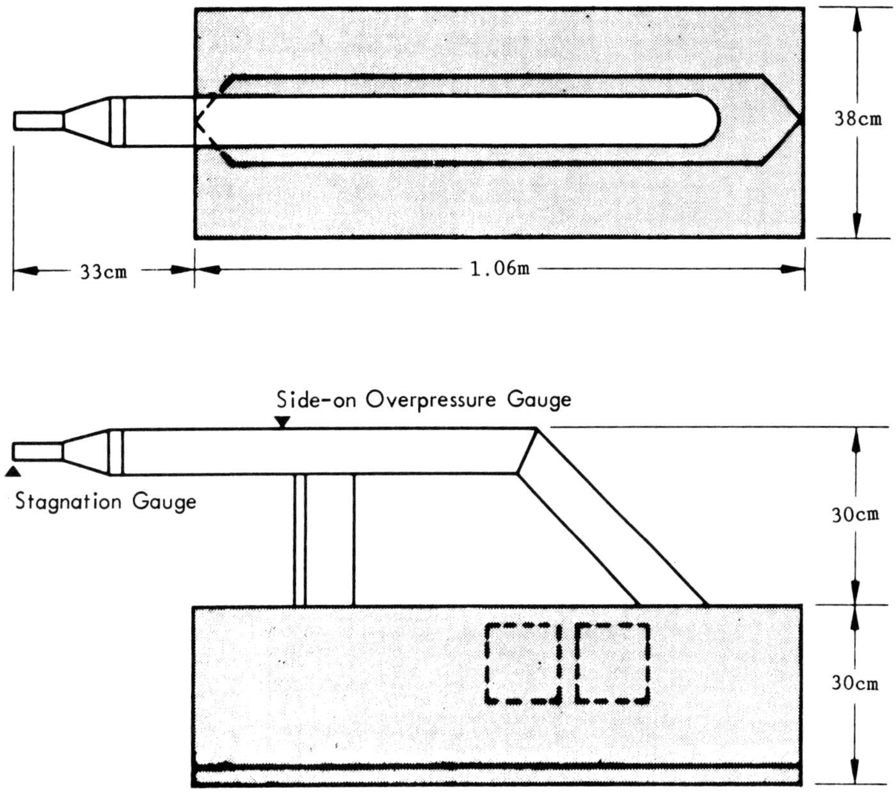

Fig. 6-11. Type C sensor mount. [Willoughby, et al (1968)]

TABLE 6-2. Summary of Blast Instrumentation for CBM & CBGS Tests [Willoughby, et al (1968)]

NO.	GAUGE LINE	NOMINAL DISTANCE (m)	90.7 kg		453 kg		11,300 kg		45,300 kg	
			P_s^*	P_o^{**}	P_s	P_o	P_s	P_o	P_s	P_o
	A			X						
I	B	1.3								
	C									
	A		X	X		X				
II	B	2.0								
	C									
	A		X	X	X	X				
III	B	3.4	X	X						
	C			X						
	A		X	X	X	X		X		X***
IV	B	5.9		X	X	X				
	C			X		X				X***
	A			X	X	X	X	X		
V	B	10		X		X				X
	C			X		X				X
	A			X		X	X	X		
VI	B	17		X		X	X	X		X
	C			X		X		X		X
	A			X		X	X	X		X
VII	B	30		X		X		X		X
	C			X		X		X		X
	A							X		X
IX	B	91						X		X
	C							X		X
	A							X		X
X	B	152						X		X
	C							X		X
	A							X		X
XI	B	270						X		X
	C							X		X

* Head-on-oriented stagnation pressure sensor.

** Side-on-oriented overpressure sensor.

*** A single gauge at the 5.9-m distance was alternately located along gauge lines A and C.

C. Blast waves from liquid propellant explosions

1. General

Explosions from liquid rocket propellant accidents "drive" air blast waves, which can in turn cause direct damage and can accelerate fragments or nearby objects. The launch pads at the Air Force Eastern Test Range (ETR) were for a number of years instrumented with air blast recorders to measure the overpressures generated during launch pad explosions, so some data were available on the intensities of the blast waves generated, even prior to Project PYRO. Such measurements, and the common practice in safety circles of comparing explosive effects on the basis of blast waves generated by TNT, have unfortunately led to expression of blast yields of propellant explosions in equivalent weight of TNT. (Although a direct conversion of weight of TNT to energy can easily be made -- 1kg of TNT equals 4520 kJ -- this has seldom been done).

Liquid propellant explosions differ from TNT explosions in a number of ways, so that the concept of "TNT equivalence" quoted in pounds of TNT is far from exact. Some of the differences are described below:

1) The specific energies (energy per unit mass) of liquid propellants, in stoichiometric mixtures, are significantly greater than for TNT (See Table 6-1).
2) Although the potential explosive yield is very high for liquid propellants, the actual yield is much lower, because propellant and oxidizer are never intimately mixed in the proper proportions before ignition.
3) Confinement of propellant and oxidizer, and subsequent effect on explosive yield, are very different for liquid propellants and TNT. Degree of confinement can seriously affect explosive yield of liquid propellants, but has only a secondary effect on detonation of TNT or any other solid explosive.
4) The geometry of the liquid propellant mixture at time of ignition can be quite different than that of the spherical or hemispherical geometry of TNT usually used for generation of controlled blast waves. The sources of compiled data for blast waves from TNT or Pentolite invariably rely on measurements of blasts from spheres or hemispheres of explosive. The liquid propellant mixture can, however, be a shallow pool of large lateral extent at time of detonation (in the CBGS simulation of launch accidents).
5) The blast waves from liquid propellant explosions show different characteristics as a function of distance from the explosion than do waves from TNT explosions. This is undoubtedly simply a manifestation of some of the differences discussed previously, but it does change the

"TNT equivalence" of a liquid propellant explosion with distance from the explosion. Fletcher (1968) discusses these differences and shows them graphically (see Figures 6-12 and 6-13). These differences are very evident in the results of the many blast experiments reported in Project PYRO. [Willoughby, et al (1968)]. They have caused the coinage of the phrase "terminal yield", meaning the yield based on blast data taken at great enough distance from the explosion for the blast waves to be similar to those produced by TNT explosions. At closer distance, two different yields are usually reported; an overpressure yield based on equivalence of side-on peak overpressures, and an impulse yield based on equivalence of side-on positive impulses.

Because of the prevalence in explosives safety circles of the TNT equivalence concepts, the predictive third volume of Willoughby, et al (1968) essentially forces a conversion of scaled liquid propellant blast data into "equivalent" weights of TNT. In later reviewing these data and techniques, however, Baker, et al (1977) saw no need for such conversion and instead presented the scaled PYRO data directly in terms of scaled distances, adjusted to agree with measured terminal yield percentages. This latter presentation, slightly modified, is used in this book.

2. Terminal yield estimation.

Of the various classes of accidental explosions discussed in this book, there are two in particular where very much less than the theoretical maximum explosive source energy or yield is released to drive an air blast wave or do other mechanical damage. These classes are UVCEs and liquid propellant explosions. In both classes, the potential for explosive yield can be great because fuels with high heats of combustion could theoretically be mixed with an oxidizer in optimum (stoichiometric or nearly so) ratio, and then ignited. But, in actuality, it is essentially impossible in accident situations for such worst-case mixing and ignition to occur. Furthermore, the most serious types of reactions - detonations - can only occur in UVCEs with certain types of flame accelerating mechanisms, and probably only occur in liquid propellant explosions in tiny cells which are well mixed within an overall heterogenous mixture.

So, for either of these classes of explosion, a crucial first question is, "What fraction or percentage of the theoretical maximum energy or yield is actually realized in a given explosion?" For UVCEs, we have unfortunately only very limited and poorly documented accident experience to answer this question. But, for liquid propellant explosions, we have fortunately the excellent data base of Project PYRO.

From this data base, the following are observations regarding blast yield:

1) Yield is quite dependent on the particular fuel and oxidizer being mixed.

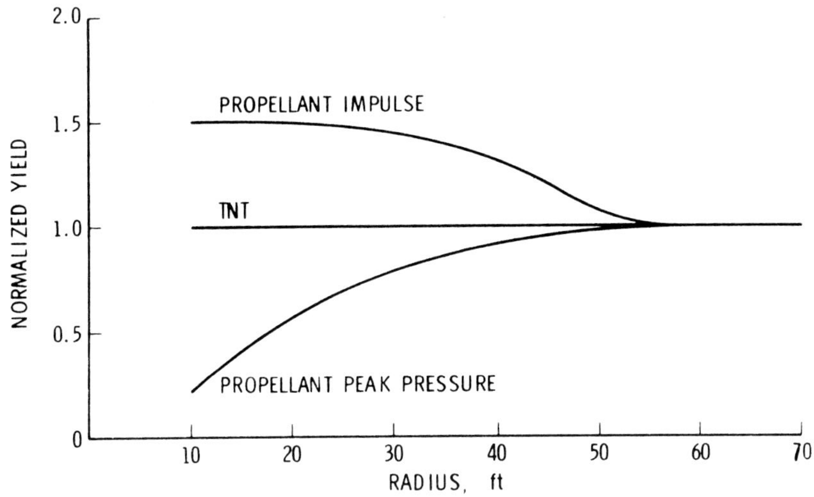

Fig. 6-12. Normalized pressure and impulse yields from explosion of N_2O_4/aerozine 50. [Fletcher (1968)]

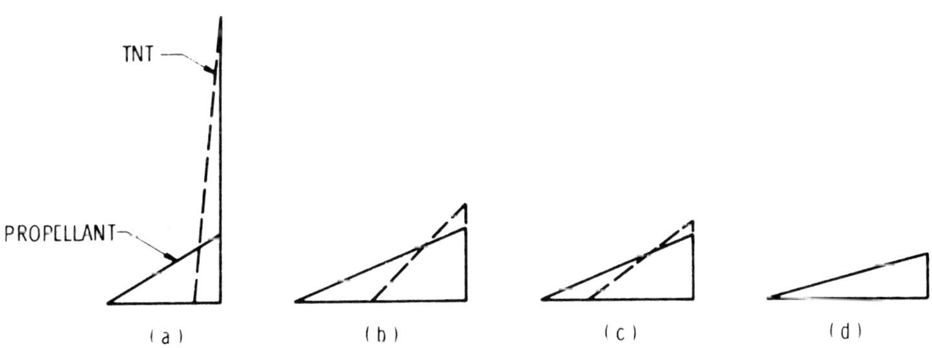

Fig. 6-13. Representative shock impulses showing coalescence of shock waves from dissimilar sources [stages (a) through (d)]. [Fletcher (1968)]

2) The yield is very dependent on the mode of mixing of fuel and oxidizer, i.e., on the type of accident which is simulated. Maximum yields are experienced when intimate mixing is accomplished before ignition.
3) On many of the liquid hydrogen/liquid oxygen (LH_2/LO_2) tests (regardless of investigators), spontaneous ignition occurred very early in the mixing process, resulting in very low percentage yields.
4) Yield is very dependent on time of ignition, even ignoring the possibility of spontaneous ignition.
5) Blast yield per unit mass of propellant decreases as total propellant mass increases.
6) Variability in yields for supposedly identical tests was great, compared to variability in blast measurements of conventional explosives.

All of the PYRO experiments, on which the prediction curves in this chapter are based, were conducted on the ground surface, with no cratering. When the curves are used to predict blast yields for explosions occurring in flight or far enough above the ground that the immediate reflection discussed does not occur, one must account for the absence of the "perfect" reflecting surface. This is done by <u>dividing</u> the blast yields calculated from curves in this chapter by a factor of <u>two</u>.

Recall that the estimates in this section are of <u>terminal yields</u>, i.e., blast yields measured far enough from a liquid propellant explosion that basic physical differences in these blast waves and those from high explosives are minor. However, we differ in our presentation in this book from that in Willoughby, et al (1968) and Baker, et al (1977) by converting the yields to <u>true</u> yields. This is intended to correct a confusing anomaly in the methods in the two earlier references wherein <u>yields greater than 100% were possible</u>, because blasts from highly energetic liquid chemicals were being compared to those from less energetic TNT. With the methods presented here, <u>true</u> yields Y_T can <u>never</u> be greater than 100%.

In general, blast yield is affected by the type of propellant, the failure mode and in some cases, ignition time, impact velocity, and type of surface impacted. Procedures for various combinations follow.

1) Hypergolic materials, by definition, ignite spontaneously on contact, so it is not possible to obtain appreciable mixing before ignition unless the fuel and oxidizer are thrown violently together. Ignition time is therefore not an important determinant of blast yield for hypergolics, but impact velocity and degree of confinement after impact are important factors. If a CBM or CBGS failure mode is being considered, percent true explosive yield can be acquired from Table 6-3. If a HVI failure mode

is assumed, then percent explosive yields can be determined from Fig. 6-14.

TABLE 6-3. Estimate of True Terminal Yields for Hyperbolic CBM and CBGS. [Willoughby, et al (1968)]

Failure Mode	Terminal Yield Range %	Estimated Upper Limit
Diaphragm rupture (CBM)	0.004 - 0.3	0.63
Spill (CBGS)	0.008 - 0.3	0.2
Small explosive donor	0.3 - 0.5	0.8
Large explosive donor	1.4 - 1.5	2
Command destruct	0.13 - 0.15	0.2
94-m drop (CBGS)	0.63	1.3

2) Because liquid oxygen/hydrocarbon propellants are not hypergolic, considerable mixing can occur in various types of accidents, and time of ignition after onset of mixing is an important determinant of blast yield. For the case of mixing and an explosion within the missile tankage (CBM), percent explosive yield can be determined by assuming an ignition time and then examining Fig. 6-15[2]. For simulated fall-back on the launch pad (CBGS), impact velocity as well as ignition time are important parameters in estimating blast yield. A two-step approach has been developed in calculating blast yield. After assuming an impact velocity, maximum percent yield Y_m can be determined for Equation (1):

$$Y_m = 4\% + 5.46\,U_I\,(\text{m/s}) \qquad (1)$$

where Y_m is expressed in percent and U_I is in meters per second. Percent explosive yield can then be determined from Y_m, an estimate of ignition

[2]A word of explanation will help clarify the meaning of the central solid line and shaded area of this graph and similar subsequent graphs. The shaded portion represents an area in which data from actual propellant blasts was found. The central solid line is an estimate of the most likely occurrence and, for most cases, is the recommended choice. Conservative estimates of explosive yield can be made by choosing the uppermost boundary of the shaded area. The vertical depth of the shaded area at any abscissa indicates the total range of data, and therefore the total uncertainty in the estimate.

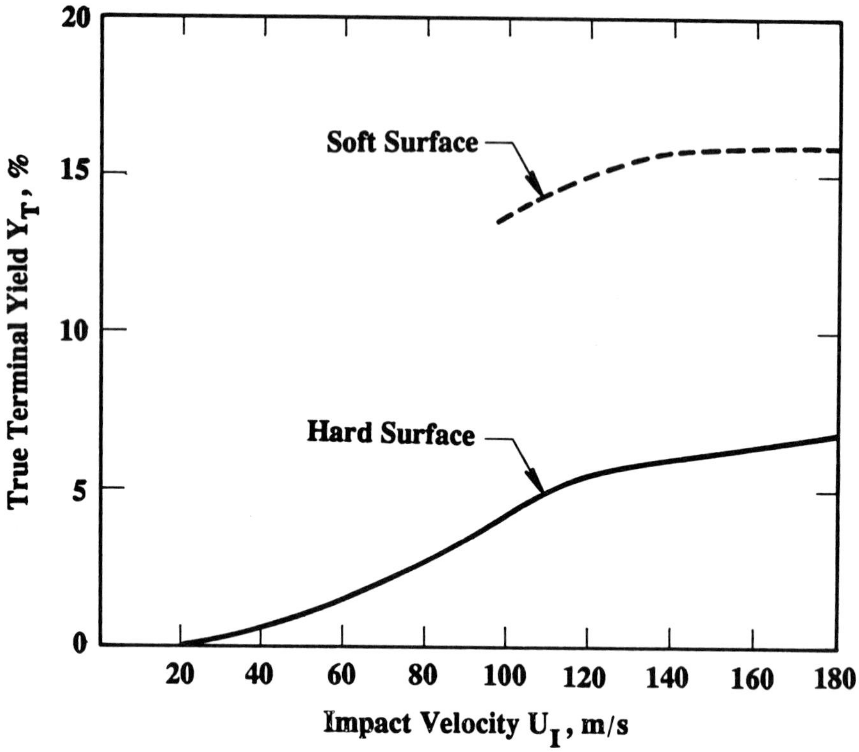

Fig. 6-14. Terminal yield vs impact velocity for hypergolic HVI. [Willoughby et al (1968)]

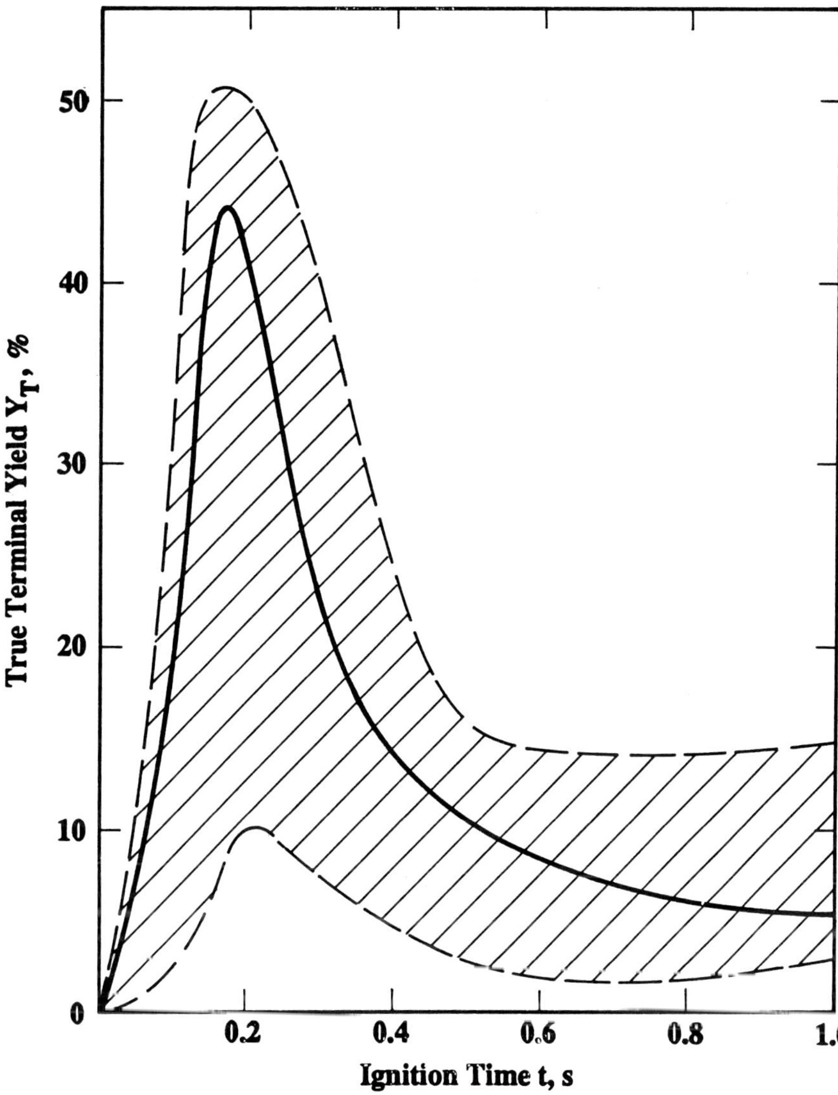

Fig. 6-15. Terminal yield vs ignition time for LO_2/RP-1 CBM. [Baker, et al (1977)]

time, and Fig. 6-16. The determination of explosive yield for the HVI failure mode is somewhat simpler because there is little ignition delay and therefore only impact velocity affects yield. Thus, blast yield can be acquired by Using Fig. 6-17 directly.

3) The determination of explosive yield, for the entirely cryogenic combination of liquid hydrogen (LH_2) fuel and liquid oxygen (LO_2) oxidizer is similar to that of liquid oxygen-hydrocarbon propellants. For the CBM case, it is necessary for one to assume an ignition time and then use Fig. 6-18 to find explosive yield. For the CBGS case, an impact velocity is assumed and maximum percent yield Y_m can be determined from Equation (2):

$$Y_m = 2.7\% + 1.20\% \, U_I \, (m/s) \tag{2}$$

where Y_m is expressed in percent and U_I is in meters per second. Percent explosive yield can then be determined from Y_m, an estimate of ignition time and Fig. 6-19. For high velocity impact (HVI) of this propellant, the blast yield is dependent only on the impact velocity and can be acquired from Fig. 6-20 directly.

3. Blast wave properties

The manner of compiling and presenting blast wave data in Project PYRO in Volumes 1 and 2 of Willoughby, et al, (1968) is based on Hopkinson-Cranz blast scaling and normalizing the data to a TNT terminal yield for each test. In the prediction methods in Volume 3 of Willoughby, et al, the TNT equivalence concept is unfortunately carried further by forcing overpressure and scaled impulse variations with scaled distance to agree with compiled TNT data. So, the basic differences in the variations of these parameters with distance for TNT and liquid propellant explosions are ignored. In Baker, et al, the authors avoided the latter trap, but still required the steps of conversion of the propellant explosion source to some equivalent weight of TNT for terminal yield, and then gave curves drawn through Hopkinson-Cranz scaled PYRO data. Here, we take the additional steps needed to convert the PYRO test data to Sachs'-scaled (dimensionless) form, before presenting scaled curves for blast wave overpressure and specific impulse. Because there are three propellants and several simulated accident situations, there are of course a number of scaled curves.

The curves presented here as Figs. 6-21 through 6-31 are dimensionless forms of Figures 2-7 through 2-17 of Baker et al (1977). The conversion equations, for SI units, are as follows:

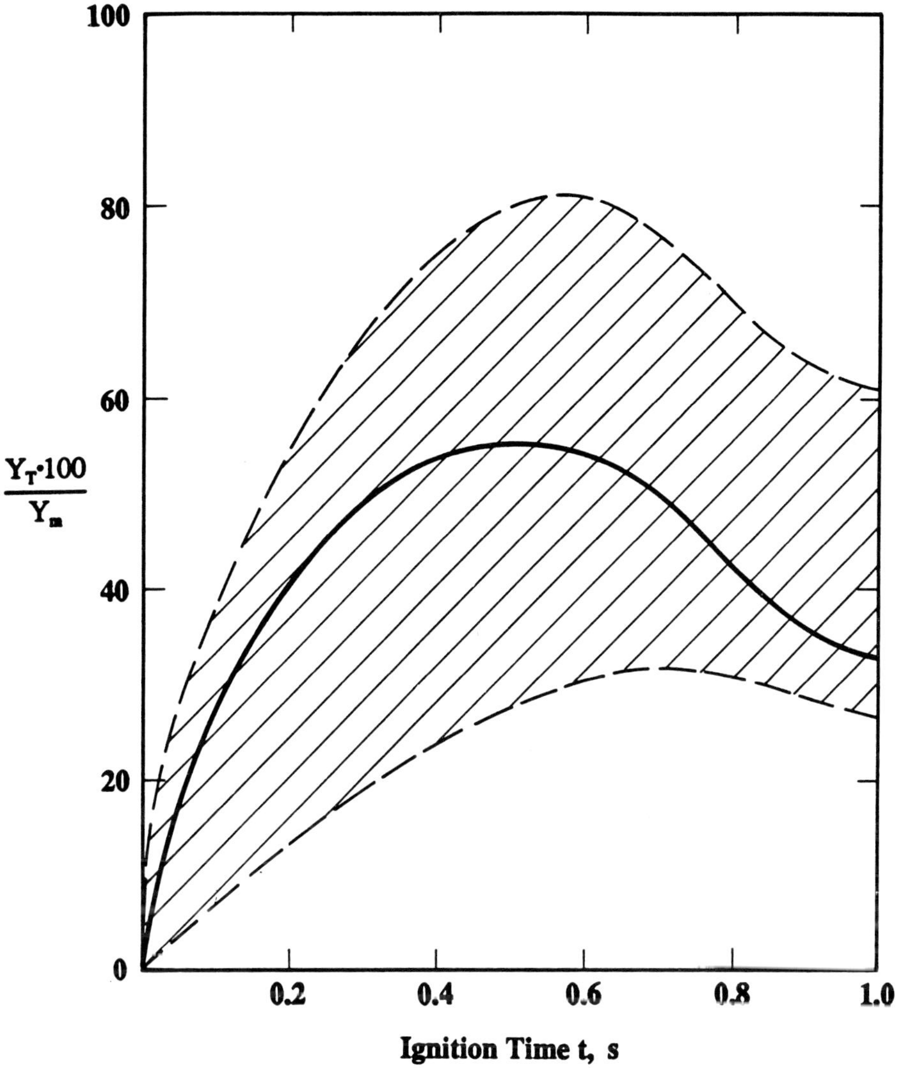

Fig. 6-16. Normalized terminal yield vs ignition time for LO_2/RP-1 CBGS. [Baker, et al (1977)]

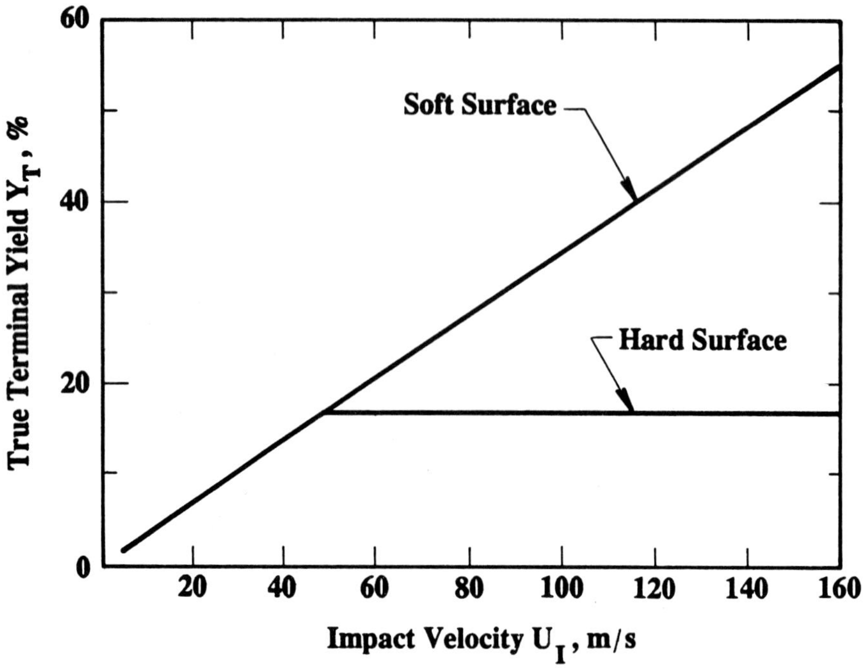

Fig. 6-17. Terminal yield vs impact velocity for LO_2/RP-1 HVI. [Willoughby, et al (1968)]

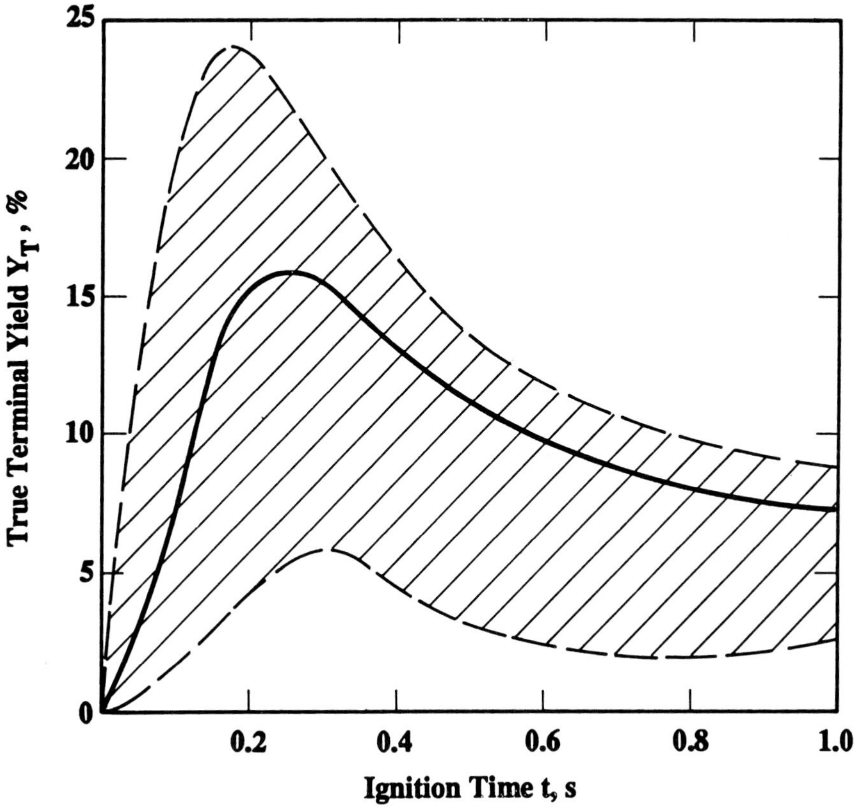

Fig. 6-18. Terminal yield vs ignition time for LO_2/LH_2 CBM. [Baker, et al (1977)]

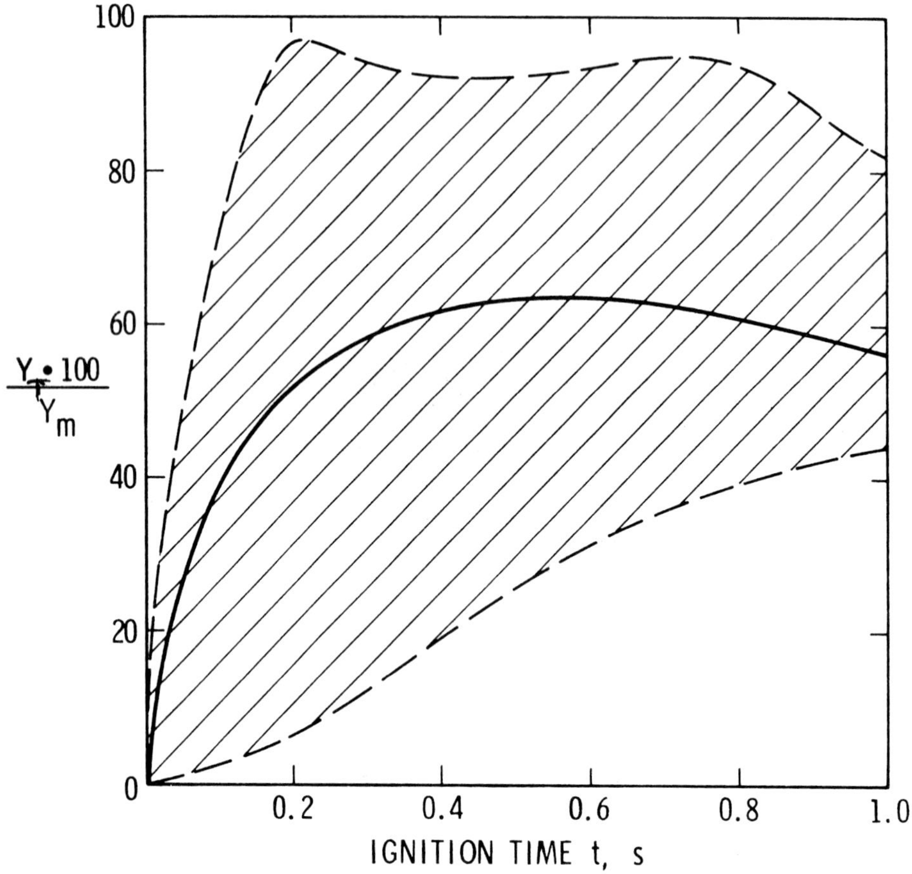

Fig. 6-19. Normalized terminal yield vs ignition time for LO_2/LH_2 CBGS. [Baker, et al (1977)]

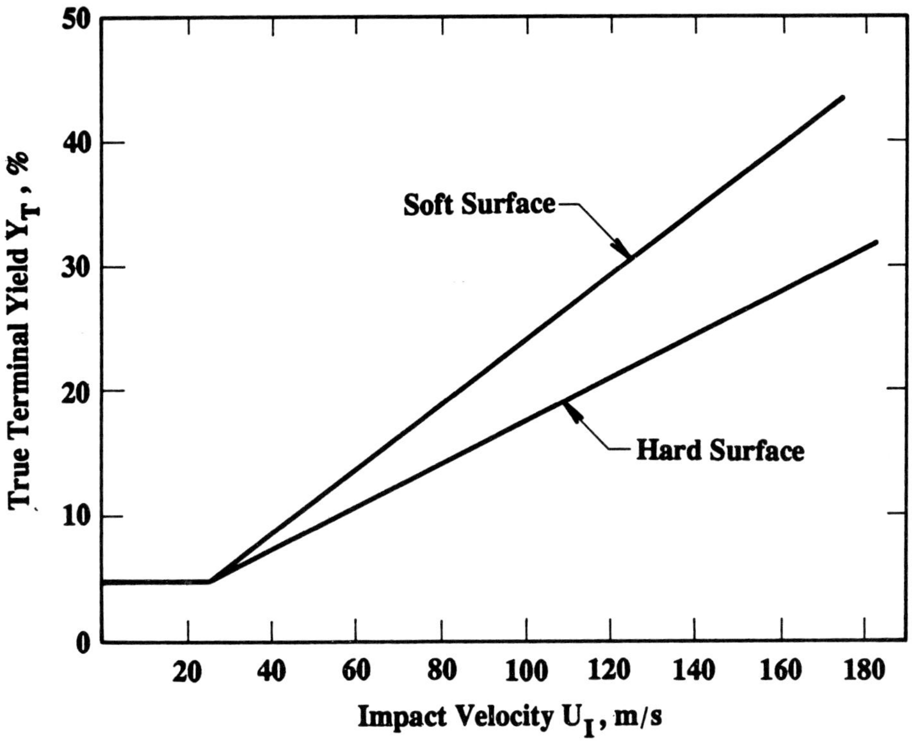

Fig. 6-20. Terminal yield vs impact velocity for LO_2/LH_2 HVI. [Willoughby, et al (1968)]

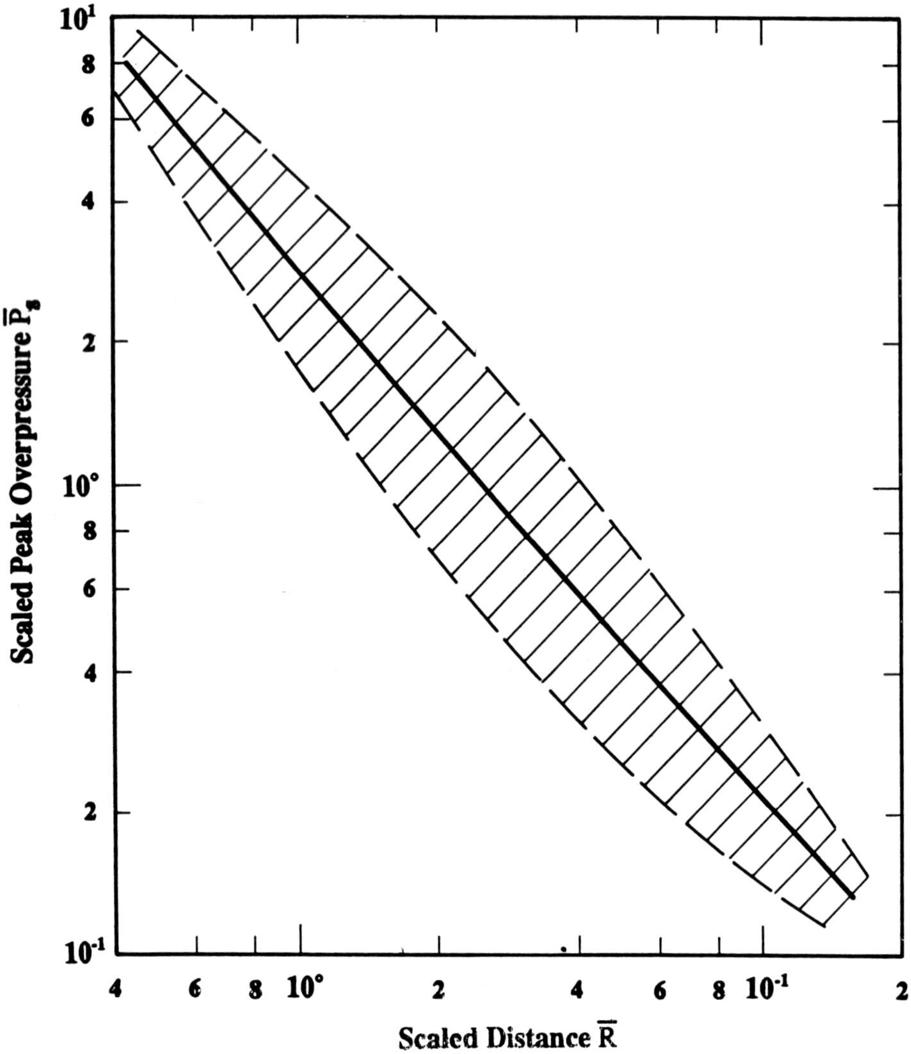

Fig. 6-21. Scaled pressure vs scaled distance. Hypergolic propellant;CBM and CBGS failure modes.

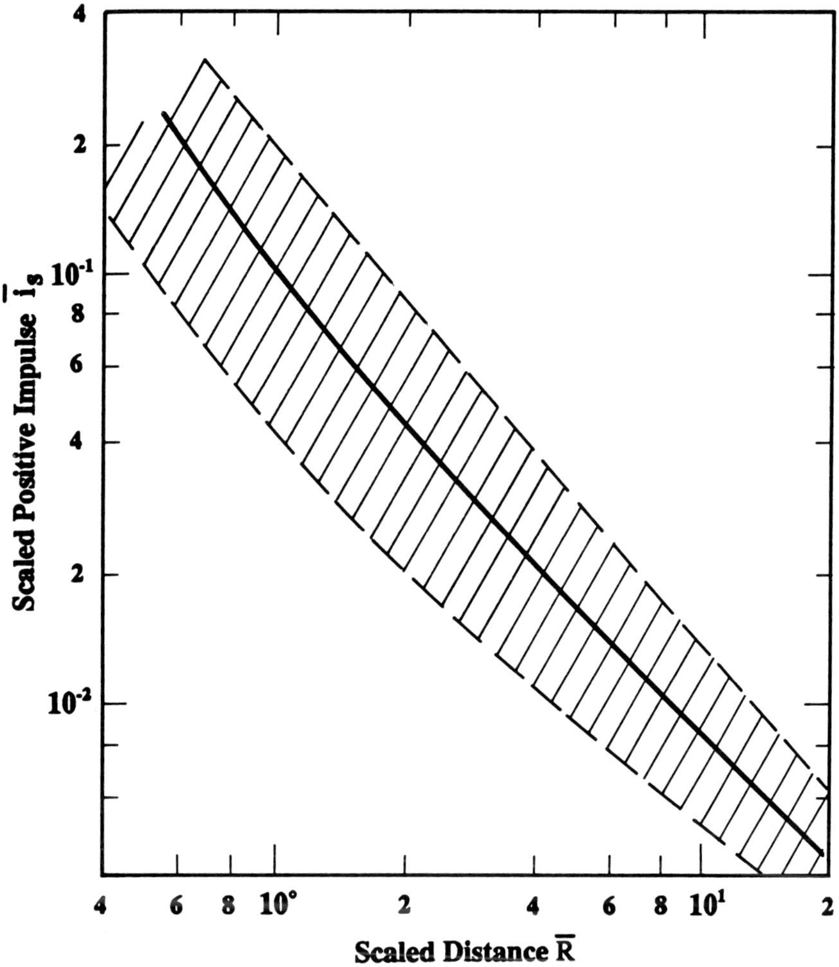

Fig. 6-22. Scaled positive impulse vs scaled distance. Hypergolic propellant; CBM, CBGS and HVI failure modes.

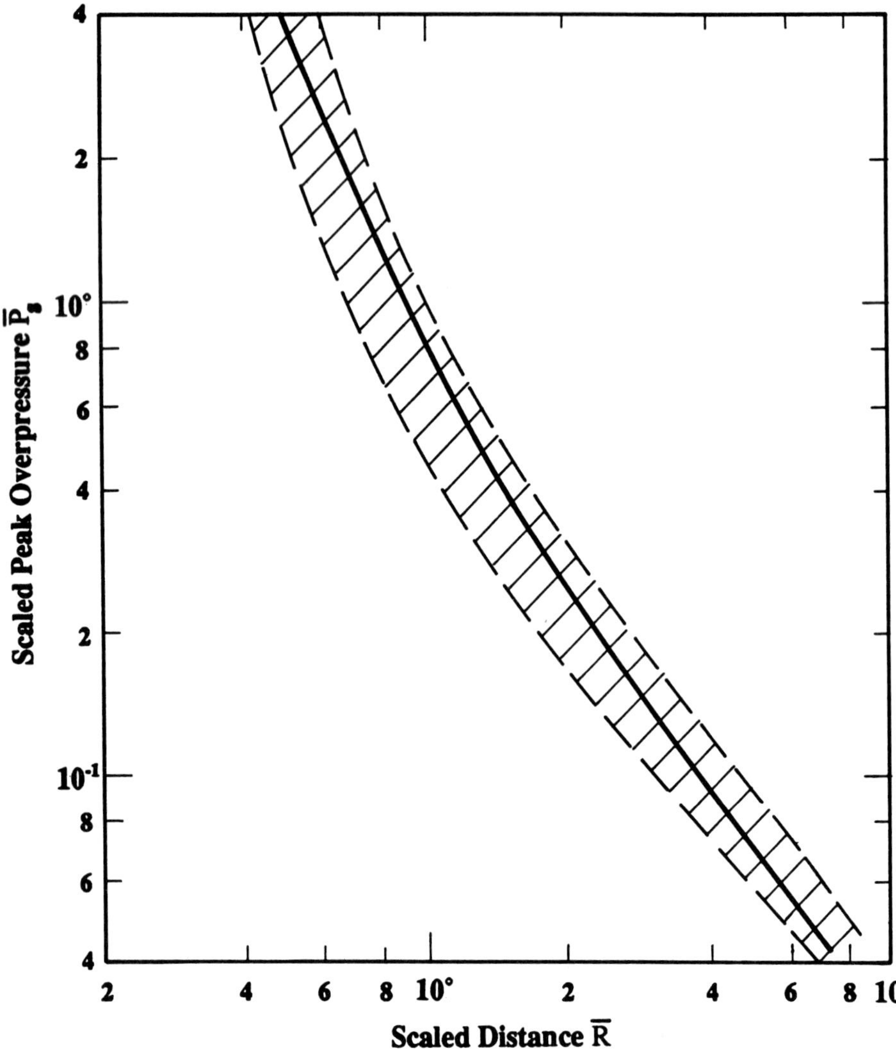

Fig. 6-23. Scaled pressure vs scaled distance. Hypergolic propellant; HVI failure mode.

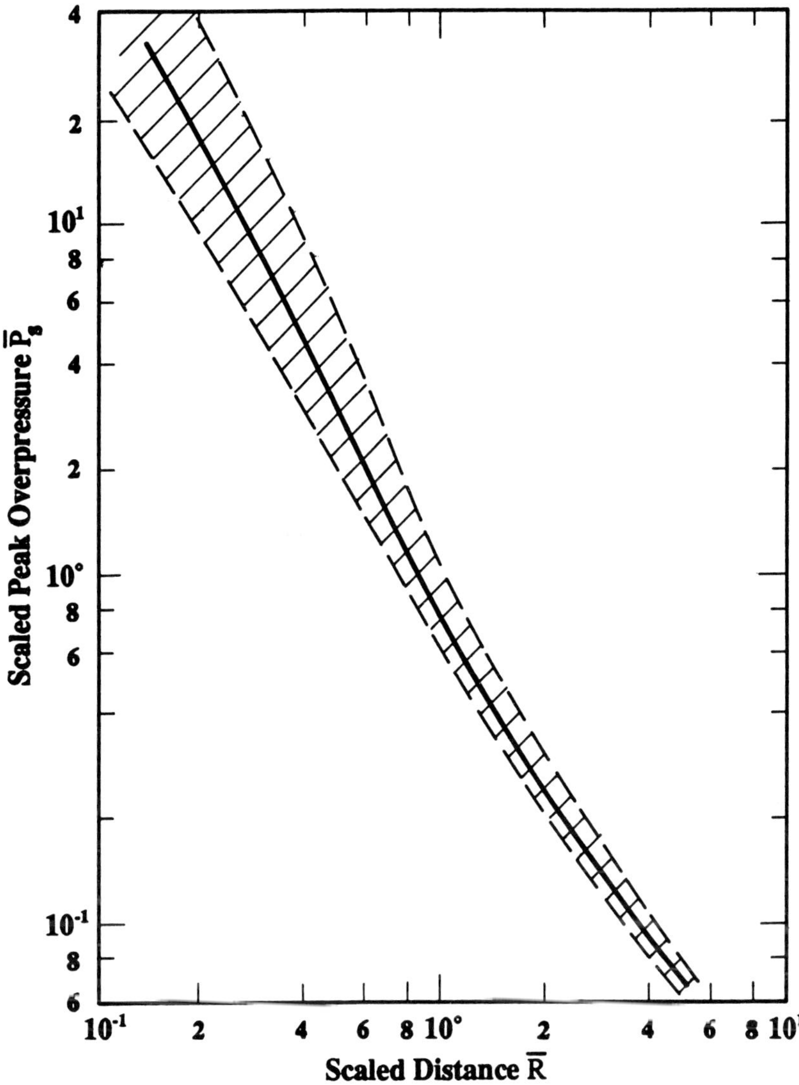

Fig. 6-24. Scaled pressure vs scaled distance. LO_2/RP-1 propellant; CBM failure mode.

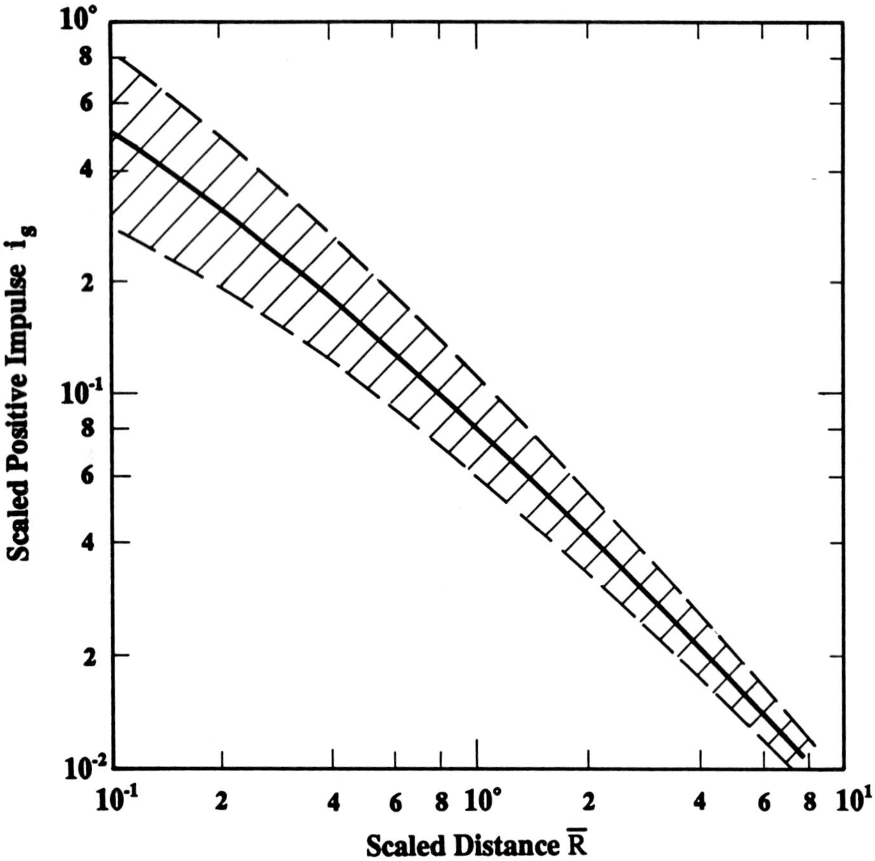

Fig. 6-25. Scaled positive impulse vs scaled distance. LO_2/RP-1 propellant; CBM failure mode.

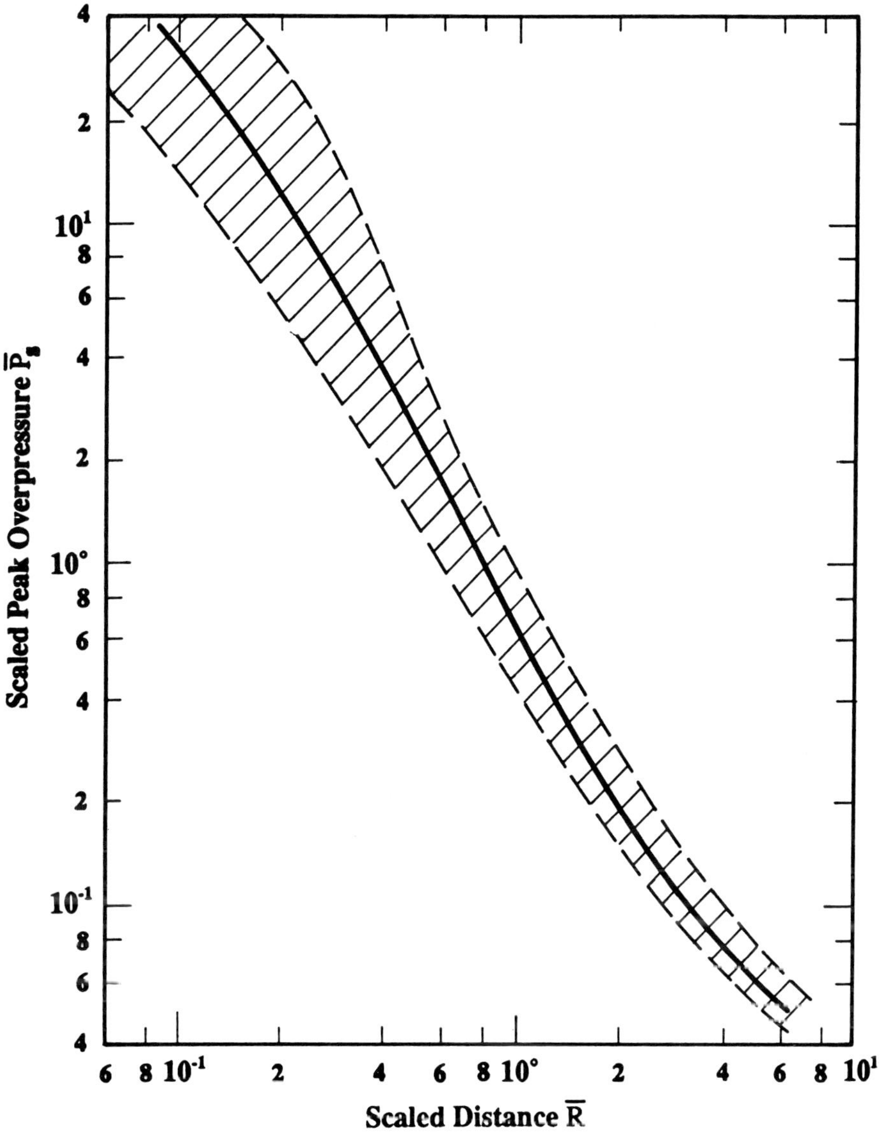

Fig. 6-26. Scaled pressure vs scaled distance. LO_2/RP-1 propellant; CBGS and HVI failure modes.

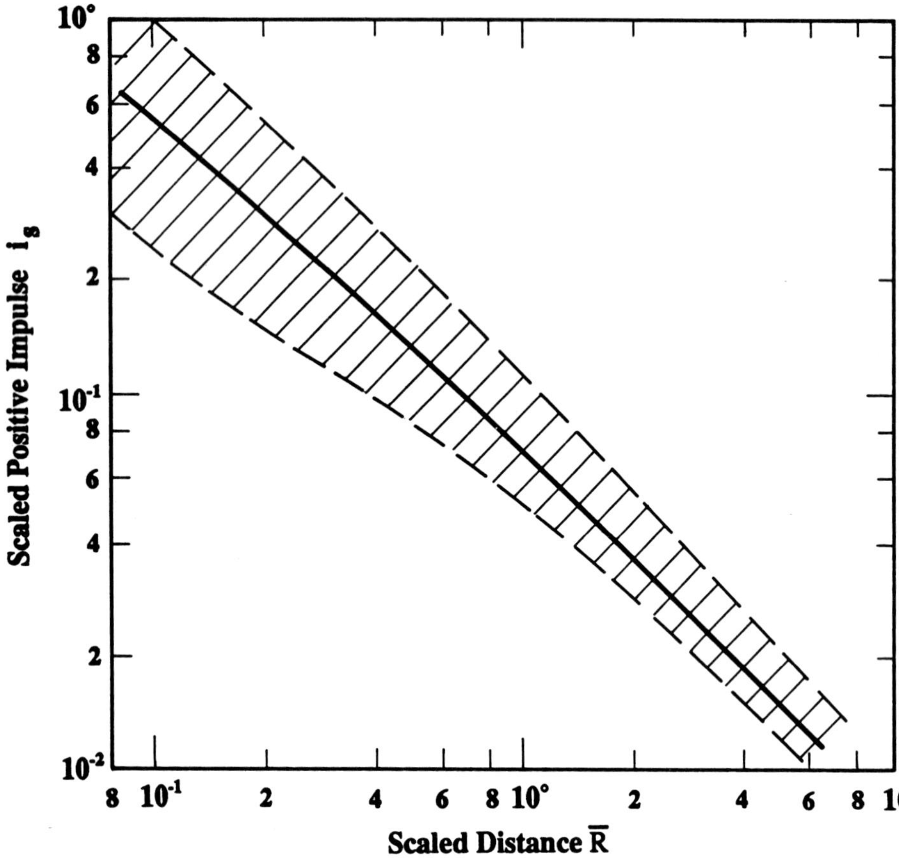

Fig. 6-27. Scaled positive impulse vs scaled distance. LO_2/RP-1 propellant; CBGS and HVI failure modes.

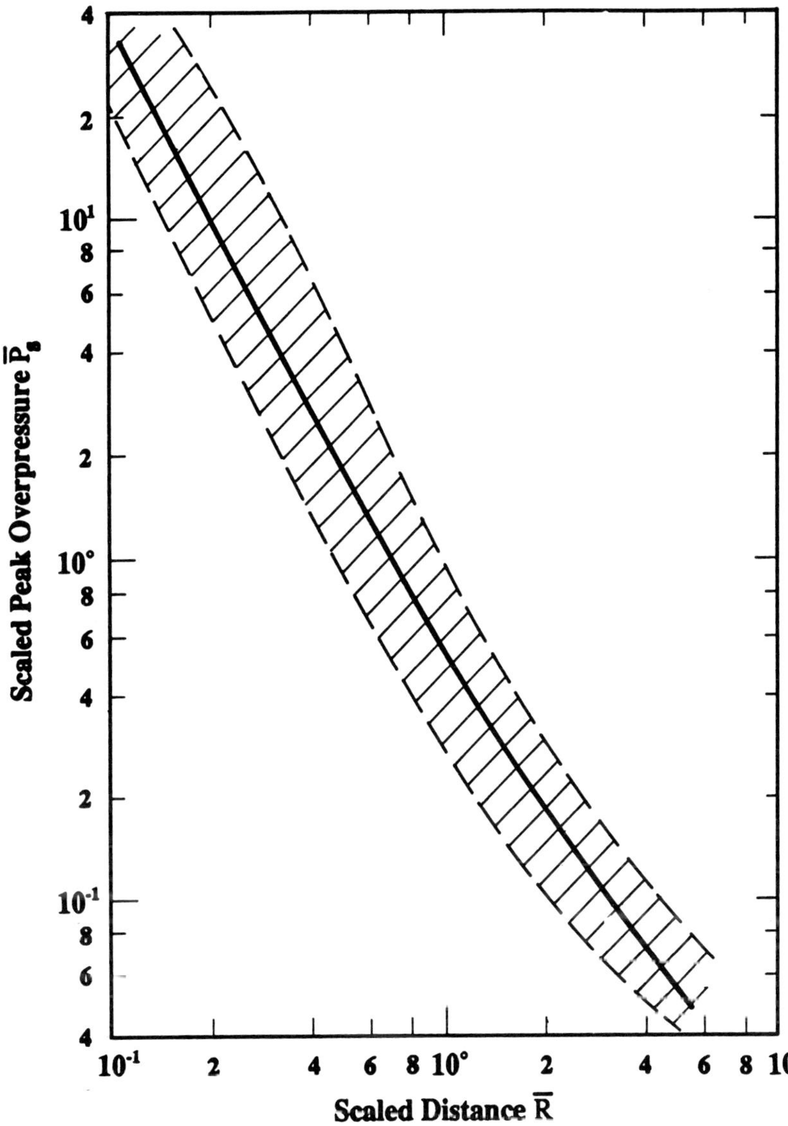

Fig. 6-28. Scaled pressure vs scaled distance. LO_2/LH_2 propellant; CBM failure mode.

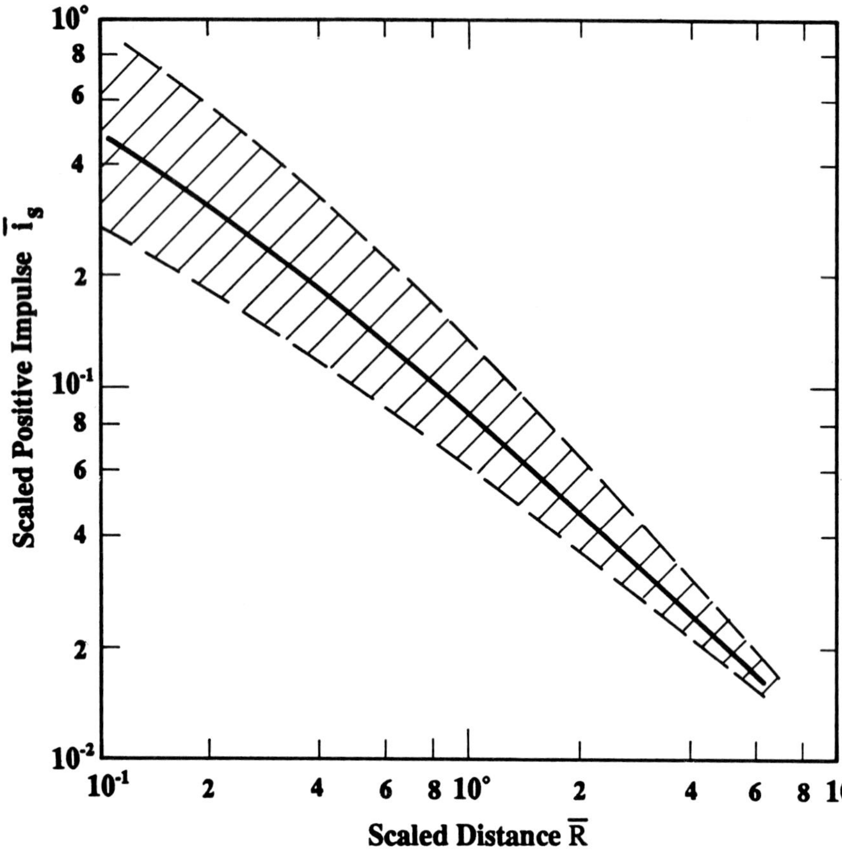

Fig. 6-29. Scaled positive impulse vs scaled distance. LO_2/LH_2 propellant; CBM failure mode.

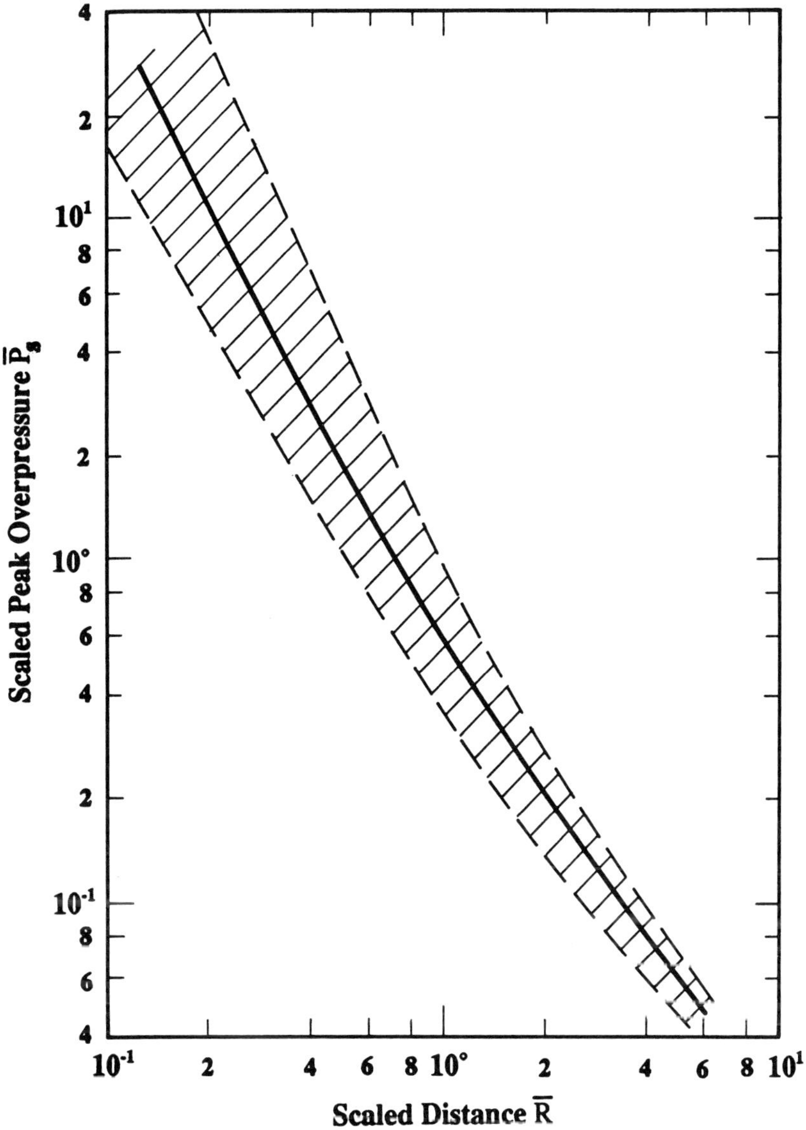

Fig. 6-30. Scaled pressure vs scaled distance. LO_2/LH_2 propellant; CBGS and HVI failure modes.

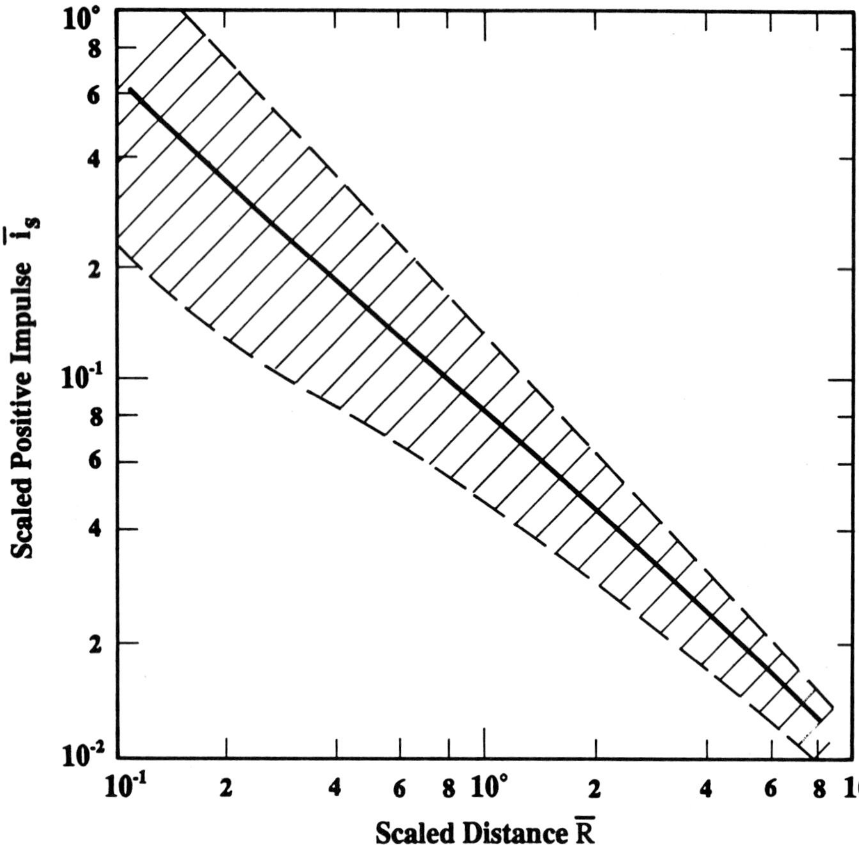

Fig. 6-31. Scaled positive impulse vs scaled distance. LO_2/LH_2 propellant; CBGS and HVI failure modes.

$$\overline{P}_s = P/p_o = \frac{P(Pa)}{1.0135 \times 10^5}$$

$$T_s = \frac{I\,a_o}{W^{1/3}\,[H_e(TNT)]^{1/3}\,(p_o)^{2/3}} \tag{3}$$

$$T_s = \frac{I \times 340\,m/s}{W^{1/3}(4520 \times 10^3 J)^{1/3}\,(1.0135 \times 10^5)^{2/3}}$$

$$T_s = (9.458\,E-4)\,\frac{I}{W^{1/3}}\frac{(Pa \cdot s)}{(kg)^{1/3}} \tag{4}$$

$$\overline{R} = \frac{R \times p_o^{1/3}}{W^{1/3} \times [H_e(TNT)]^{1/3}}$$

$$\overline{R} = \frac{R \times (1.0135 \times 10^5)^{1/3}}{W^{1/3} \times (4520 \times 10^3)^{1/3}}$$

$$\overline{R} = \frac{0.282\,R(m)}{W^{1/3}\,(kg)^{1/3}} \tag{5}$$

Please recall again, though, that these curves all relate to the surface burst conditions for Project PYRO, and that blast source energies or yields must be adjusted to account for the nearly perfect ground reflections which occur for all liquid propellant explosions if explosions occur high in the air, by division by two.

D. Closure

As opposed to HE or bursting vessel explosions, liquid propellant explosions are much more variable and ill-defined. This is true even for controlled explosion experiments, such as in Project PYRO [Willoughby, et al (1968)], and even more true for accidental liquid propellant explosions.

Perhaps the most important difference in high explosives and liquid propellants is that high explosives are "mass detonating", while liquid propellants are not. This means that all of a high explosive source detonates and affects its surroundings, if properly initiated; while liquid propellants can never be completely intimately mixed and then detonated. (Although it is very likely that small "cells" of mixed propellants do indeed detonate in liquid propellant explosions, the complete mixing and detonation of large masses is impossible because of temperature differences, phase changes, etc.) This difference can be stated in another way. High explosive sources have a high energy density, or energy per unit volume or mass, while liquid propellant sources have a low energy density.

Another large uncertainty in liquid propellant explosions is in the geometry of the mixed region of propellants. This is very dependent on the test or accident sequence and geometry. In the two most common test geometries, (simulating certain types of accidents), the propellant mixing region could be roughly globular or cylindrical (CBM case), or a thin, nearly flat sheet (CBGS case). In real accident situations, the geometry is almost impossible to predict. On the other hand, the geometry of a high explosive or bursting vessel source is known exactly prior to its detonation or explosion.

Another unknown in liquid propellant explosions is the reaction rate. It must be quite variable within a region of mixed propellant, because mixture ratios must vary widely within this region, and it is essentially certain to be, on the average, well below a detonation velocity. Not even average reaction rates have been measured in any but microscale experiments. But, high explosives have a known, essentially constant reaction rate - the detonation velocity.

Physical states are also quite different for high explosives and "mixtures" of cryogenic liquid propellants. The high explosives are either solid, or slurries with the liquid and solid components of the slurries being well intermixed and at the same temperature. Liquid hydrogen and liquid oxygen thrown together will be in violent agitation, with the hydrogen boiling and the oxygen freezing. The "mixture" will be very dynamic and heterogeneous, and include gases, liquids and solids.

Finally, another significant difference is that liquid propellants have a much higher specific energy than high explosives. That is, if they could be intimately mixed before exploding, they would release much more energy per unit mass than high explosives. For example, a stoichiometric mixture of liquid oxygen and liquid hydrogen would theoretically release about 3.7 times as much energy as an equal mass of TNT.

All of the factors cited here cause the characteristics of blast waves generated by liquid propellant explosions to be much less predictable than for HE explosions, and also of quite different character, particularly close to the explosion source. We might add that an upper limit to side-on blast overpressure of liquid propellant explosions, even for measurements within the mixed propellants, seems to be [from data in Willoughby, et al (1968)] about 14MPa.

Perhaps the reader may ask whether there are indeed later useful data or methods for helping in prediction of blast wave properties for accidental explosions in more modern liquid-fueled launch vehicles than those of the 1968 era. A recent review of explosion hazards for Space Shuttle launches (NUS-4543) uncovered no significant new test data, but much more advanced computer capability for prediction of propellant mixing and explosions.

One can predict air blast wave characteristics for certain classes of explosion sources with hydrocarbons. Generally, these codes (there are a large number of versions available) numerically solve the complex differential equations describing the generation of blast waves and their transmission through various media. They are designed to predict the entire transient pressure, temperature, and flow fields within and about detonating high explosives and other well-defined explosion sources. The proven codes of this class have been well "calibrated" against test data, but their strength lies in their ability to predict blast properties in regimes where measurements are difficult or impossible, and their ability to predict properties for which no suitable measurement techniques exist. Generally, one-dimensional and two-dimensional hydrocodes will run on a medium-sized scientific computer. But, three-dimensional codes require so much memory that they will only run on supercomputers, and are very expensive and time-consuming to run.

For HE explosions, particularly for symmetric explosion sources, hydrocodes give reasonably accurate to very accurate predictions of blast wave properties. But, for liquid propellant explosions, predictions are much less accurate. The basic reasons for these inaccuracies lie in the lack of symmetry of liquid propellant explosions, and lack of homogeneity of the propellant mixture, neither of which can be simulated by the hydrocodes.

As an example, the one dimensional hydrocode KOVEC modeled a CBM explosion as an instantaneous burn of a cylindrical annular volume of stoichiometric liquid hydrogen and solid oxygen, with a pressure of 18.6 kb (see NUS-4543). Although such pressures may exist in microscopic cells which have become well mixed, it is very unlikely that they will exist throughout a significant volume of the propellant mixture.

By contrast, the much better-defined detonation processes for high explosives, with detonation pressures in the range of hundreds of kilobars behind the detonation fronts, are very well modeled by the hydrocodes.

In general, we conclude that real life liquid propellant explosions have the following properties:

- o They are very ill-conditioned.
- o They have high specific energy, but low overall energy density.
- o They are almost always geometry- or scenario- dependent.

- o They produce low peak blast overpressures and high specific impulses close-in.
- o They are difficult to predict with hydrocode calculations.
- o They only correlate well with HE explosions in the far field.

BIBLIOGRAPHY

G. E. Abouseif and T. Y. Toong, "On Direct Initiation of Gaseous Detonations", Combustion & Flame, 45 (1982) pp. 39-46.

G. E. Andrews and D. Bradley, "Determination of Burning Velocities: A Critical Review", Combustion & Flame, 18 (1972) pp. 1-33.

Anonymous, "An Investigation of Hazards Associated with the Storage and Handling of Liquid Hydrogen, Final Report", C-61092, Arthur D. Little, Inc., Cambridge, MA, (1960a).

Anonymous, "Research on the Hazard Classification of New Liquid Rocket Propellants", Quarterly Progress Report, R-2452-1, North American Aviation, Inc., Rocketdyne Division, Canoga Park, CA, (1960b).

Anonymous, "Research on the Hazard Classification of New Liquid Propellants", Quarterly Progress Report, R-2452-3, North American Aviation, Inc., Rocketdyne Division, Canoga Park, CA, (1960c).

Anonymous, "The Handling and Storage of Liquid Propellants, Medical Services", AFM 160-39, Department of the Air Force, Washington, (1961a).

Anonymous, "Electrostatic Hazards Associated with the Transfer and Storage of Liquid Hydrogen", Final Report, C-61092, Arthur D. Little, Inc., Cambridge, MA, (1961b).

Anonymous, "Research on Hazard Classification of New Liquid Rocket Propellants, Final Report, Volume I", R-3217, Rocketdyne, A Division of North American Aviation, Inc., Canoga Park, CA, (1961c).

Anonymous, "Summary Report on a Study of the Blast Effect of the Saturn Vehicle", C-63850, Arthur D. Little, Inc., Cambridge, MA, (1962).

Anonymous, "Saturn S-II Propellant Hazards Suppression Program", Final Report, R-6081, North American Aviation, Inc., Rocketdyne Division, Canoga Park, CA, (1965a).

Anonymous, "Feasibility Study for Assessment of Explosion Hazards", Final Technical Report, General Electric Co., Missile and Space Division, Daytona Beach, FL, (1965b).

Anonymous, "Prevention of and Protection Against Accidental Explosion of Munitions, Fuels and Other Hazardous Mixtures", Annals of the New York Academy of Sciences, 152 (1968).

Anonymous, "Chemical Rocket/Propellant Hazards, Volume III, Liquid Propellant Handling and Storage and Transportation", CPIA Pub. No. 194, Chemical Propulsion Information Agency, The John Hopkins University, Applied Physics Laboratory, Silver Spring, MD, (1970).

Anonymous, "Suppressive Shields Structural Design and Analysis Handbook", HNDM-1110-1-2, U. S. Army CoE, Huntsville Div., Huntsville, AL, 1977.

Anonymous, "Guide for Explosion Venting", NFPA 68-1978, Nat. Fire Prot. Assoc., Boston, MA, 1978.

Anonymous, "The Investigation of Grain Elevator Explosions, Publication NMAB 367-1", National Academy Press, Washington, DC, 1980a.

Anonymous, Methods for the Calculation of the Physical Effects of the Escape of Dangerous Material (Liquids and Gases), Part I and II, Bureau for Ind. Safety TNO, Voorburg, Neth., 1980b.

Anonymous, "Chemical Data Guide for Bulk Shipment by Water", CIM 16616.6, U. S. Coast Guard, Washington, DC, 1982a.

Anonymous, "Prevention of Grain Elevator and Mill Explosions", Report of the Panel on Causes and Prevention of Grain Elevator Explosions of the Comm. on Eval. of Ind. Hazards, National Academy Press, Washington, DC, 1982b.

Anonymous, "Estimating Air Blast Characteristics for Single Point Explosions in Air, with a Guide to Evaluation of Atmospheric Propagation Effects", ANSI S2.20-1983 (ASA 20-1983), Acous. Soc. of Amer., New York, NY, 1983.

Anonymous, "First International Colloquium on Explosibility of Industrial Dusts", Baranow, Poland, 1984a.

Anonymous, The Control and Prevention of Gas Explosions, Oyez Sci. & Tech. Serv., Ltd, London, England, 1984b.

Anonymous, "Report of Shuttle Transportation System (STS) Explosion Working Group (EWG)", NUS-4543 (Revised), for Dept. of Energy, NUS Corp., Gaithersburg, MD, 1985.

Anonymous, "NFPA 69. Explosion Protection Systems, 1986", Nat. Fire Prot. Assoc., Quincy, MA, 1986.

Anonymous, "Shengyang International Symposium on Dust Explosions", Shengyang, China, 1987.

Anonymous, "NFPA 68. Venting of Deflagrations. 1988 Edition", Nat. Fire Prot. Assoc., Quincy, MA, 1988a.

Anonymous, Flammable Dust Explosions, Int. Bus. Comm., St. Louis, MO, 1988b.

Anonymous, "NFPA 70. National Electrical Code", Nat. Fire Prot. Assoc., Quincy, MA, 1990.

E. J. Anthony, "Some Aspects of Unconfined Vapor Cloud Explosions", Jour. Haz. Mat., 1 (1976) pp. 289-301.

H. Axelsson and S. Berglund, "Cloud Development and Blast Wave Measurements from Detonation Fuel-Air Explosive Charges", FOA Report C 20225-D4, FOA, Stockholm, Sweden, 1978.

G. Bach, R. Knystautas and J. H. S. Lee, "", 13th Symp. Int. Combustion Proc., 1970.

G. G. Bach and J. H. S. Lee, "An Analytical Solution for Blast Waves", AIAA Journal, 8 (1970) pp. 271-275.

B. E. Bader, A. B. Donaldson and H. C. Hardee, "Liquid-Propellant Rocket Fire Abort Model", J. Spacecraft, 8 (1971).

B. E. Bader, et al, "Liquid-Propellant Rocket Fire Abort #13 Contract Model", SC-RR-70-454, UC-36, (1976).

W. E. Baker, Explosions in Air, Wilfred Baker Engineering, San Antonio, TX, 1983.

W. E. Baker, "Blast Waves Emitted from Gas Explosions", The Control and Prevention of Gas Explosions, Oyez Sci. & Tech. Services Ltd., London, England, 1984.

W. E. Baker, Explosion Prevention, Audio Course C-95, Am. Chem. Soc., Washington, DC, 1987.

W. E. Baker, P. A. Cox, P. S. Westine, J. J. Kulesz and R. A. Strehlow, Explosion Hazards and Evaluation, Elsevier Scientific Publishing Co., Amsterdam, 1983.

W. E. Baker, E. D. Esparza, J. C. Hokanson, J. E. Funnel, P. K. Moseley and D. M. Deffenbaugh, "Initial Feasibility Study of Water Vessels for Arresting Lava Flow", AMSAA FEAT Interim Note No. F-13, AMSAA, Aberdeen, MD, 1978.

W. E. Baker, E. D. Esparza and J. J. Kulesz, "Venting of Chemical Explosions and Reactions", Proceedings of 2nd International Symposium on Loss Prevention and Safety Promotion in the Process Industries, Heidelberg, Germany, (1977).

W. E. Baker, J. C. Hokanson and J. J. Kulesz, "A Model Analysis for Vented Dust Explosions", Proceedings of the 3rd International Symposium on Loss Prevention and Safety Promotion in the Process Industries, Basel, Switzerland, (1980) pp. 1339-1347.

W. E. Baker, J. J. Kulesz, R. E. Ricker, R. L. Bessey, P. S. Westine, V. B. Parr and G. A. Oldham, "Workbook for Predicting Pressure Wave and Fragment Effects of Exploding Propellant Tanks and Gas Storage Vessels", NASA CR-134906, NASA Lewis Research Center, (1975).

W. E. Baker, J. J. Kulesz, R. E. Ricker, P. S. Westine, V. B. Parr, L. M. Vargas and P. K. Moseley, "Workbook for Estimating Effects of Accidental Explosions in Propellant Ground Handling and Transport Systems", National Aeronautics and Space Administration, Lewis Research Center, NASA Contractor Report 3023, (1978).

W. E. Baker, V. B. Parr, R. L. Bessey and P. A. Cox, "Assembly and Analysis of Fragmentation Data for Liquid Propellant Vessels", NASA, Lewis Research Center, Aerospace Safety Research and Data Institute, Cleveland, OH, NASA CR-134538., (1974).

W. E. Baker, P. S. Westine and F. T. Dodge, Similarity Methods in Engineering Dynamics, The Hayden Book Company, Rochelle Park, NJ, 1973.

W. Bartknecht, Explosionen: Ablauf und Schutzmassnahmen, Springer-Verlag, Berlin, Heidelberg, 1978a.

W. Bartknecht, "Gas Vapor and Dust Explosions", Presented at the International Symposium on Grain Elevator Explosions, Washington D.C., (1978b).

W. Bartknecht, Explosions - Course, Prevention, Control, Springer-Verlag, Heidelberg, Germany, 1981.

W. Bartknecht, "Preventive and Design Measures for Protection Against Dust Explosions", Industrial Dust Explosions, ASTM STP 958, ASTM, Philadelphia, PA, 1986.

M. R. Baum, "Blast Waves Generated by the Rupture of Gas Pressurized Ductile Pipes", Trans. I. Chem. E., 57 (1979) pp. 15-24.

W. B. Benedick, "High Explosive Initiation of Methane-Air Detonations", Combustion and Flame, 35 (1979) pp. 89-94.

W. B. Benedick, "Private communication", 1983.

W. B. Benedick, C. M. Guirao, R. Knystautas and J. H. S. Lee, "Crirical Charge for the Direct Initiation of Detonation in Gaseous Fuel-Air Mixtures", Dynamics of Explosions, AIAA, New York, NY, 1986.

C. V. Bishop, F. J. Benz and L. J. Ullian, "Mixing of Cryogenic Fluids and Predicted Detonation Properties for Multiphase Liquid Oxygen and Liquid Hydrogen", 1986 JANNAF Propulsion Meeting, 1986.

D. R. Blackmore, J. A. Eyre and G. G. Summers, "Dispersion and Combustion Behaviour of Gas Clouds Resulting from Large Spills of LNG and LPG onto the Sea", Trans. of the Inst. of Marine Eng., 94 (1982).

C. H. Blakewood, et al, "Preliminary Investigation of Blast Hazards of RP-1/LOX and LH-2-1/LOX Propellant Combinations", NASA, Marshall Space Flight Center, Huntsville, AL, NASA-TM-X-53240, (1965).

F. T. Bodurtha, Industrial Explosion Prevention and Protection, McGraw-Hill Book Company, New York, 1980.

A. A. Boni, M. Chapman and J. L. Cook, 6th ICDERS, 1977.

A. A. Borisov and S. A. Loban, "", Fiz Goreniya Vzryva, 13 () p. 929.

D. W. Boyer, H. L. Brode, I. I. Glass and J. G. Hall, "Blast from a Pressurized Sphere", UTIA Report No. 48, Toronto, Canada, 1958.

F. V. Bracco, "Air Shock Parameters and Design Criteria for Rocket Explosions", WR 65-21, Wyle Laboratories, Testing Division, Huntsville Facility, Huntsville, AL, (1965a).

F. V. Bracco, "A Method for Predicting the Air Blast Parameters from Liquid Propellant Rocket Explosions", WR-64-11, Wyle Laboratories, Testing Division, Huntsville Facility, Huntsville, AL, (1965b).

F. V. Bracco, "Preliminary Results of the Calculation of Air Shock Parameters in the Close-Field of Liquid Propellant Explosions", WR-66-18, Wyle Laboratories, Testing Division, Huntsville Facility, Huntsville, AL, (1966).

D. Bradley, S. El-Din Habik and M. A. M. Jamel, "The Roles of Alumina and Graphite Dust Additives in Methane-Air Flames", First Int. Coll. on Explosibility of Industrial Dusts, Baranow, Poland, 1984.

D. Bradley and A. Mitcheson, "The Venting of Gaseous Explosions in Spherical Vessels, Parts I & II", Combustion & Flame, 32 (1978) pp. 221-255.

S. R. Brinkley, Jr., "Determination of Explosion Yield of an Exothermic or Detonable Reaction", Symposium on Loss Prevention in the Process Industries, Part V, PrePrint No. 29(b), New Orleans, LA, (1969).

H. L. Brode, "The Blast from a Sphere of High Pressure Gas", Report P-582, Rand Corp., 1955.

H. L. Brode, "The Blast Wave in Air Resulting from a High Temperature, High Pressure Sphere of Air", Report RM-1825-AEC, Rand Corp., 1956.

H. L. Brode, "Quick Estimates of Peak Overpressure from Two Simultaneous Blast Waves", Defense Nuclear Agency Report No. DNA4503T, Contract No. DNA001-78-C-0009, R&D Assoc., RDA-TR-107006-008, Marina Del Rey, CA, (1977).

J. Brossard, J. C. Leyer, D. Desbordes, J. P. Saint-Cloud, S. Hendricks, J. L. Garnier, A. Lannoy and J. L. Perrot, "Air Blast from Unconfined Gaseous Detonations", Dynamics of Shock Waves, Explosions and Detonations, AIAA, New York, NY, 1984.

S. J. Brown, ed., Impact, Fragmentation and Blast. (Vessels, Pipes, Tubes, Equipment), American Society of Mech Engr, New York, 1984.

D. C. Bull and J. E. Elsworth, "Concentration Limits to Unconfined Detonations of Ethane/Air", Combustion & Flame, (1978).

D. C. Bull, J. E. Elsworth and G. Hooper, "Initiation of Spherical Detonation in Hydrocarbon-Air Mixtures", Acta Astronautica, 5 (1978) pp. 997-1008.

D. C. Bull, J. E. Elsworth, P. J. Shuff and E. Metcalfe, "Detonation Cell Structures in Fuel/Air Mixtures", Combustion & Flame, 45 (1982) pp. 7-22.

D. C. Bull and J. A. Martin, "Explosion of Unconfined Clouds of Natural Gas", AGA Transmission Conference, St. Louis, MO, 1977.

D. Burgess, J. Grumer and M. G. Zabetakis, "Research on the Fire and Explosion Hazards Research Associated with New Liquid Propellants", Annual Report, Bureau of Mines, U.S. Department of the Interior, Pittsburgh, PA, (1960a).

D. Burgess, J. Grumer and M. G. Zabetakis, "Research on the Fire and Explosion Hazards Associated with New Liquid Propellants", Progress Report No. 3, Bureau of Mines, U.S. Department of the Interior, (1960b).

D. Burgess, J. Grumer and M. G. Zabetakis, "Research on the Fire and Explosion Hazards Associated with New Liquid Propellants", Progress Report No. 8, Bureau of Mines, U.S. Department of the Interior, Pittsburgh, PA, (1961).

J. Campos, A. R. Janeiro Borges and R. Bouriannes, "Influence of Heat Transfer on the Explosion Pressure of Lean Mixtures of Methane-Air-Graphite", Dynamics of Explosions, AIAA, Newe York, NY, 1986.

K. L. Cashdollar and M. Hertzberg, Industrial Dust Explosions, Symposium on Industrial Dust Explosions, Am. Soc. for Test. & Mats., Pittsburgh, PA, 1987.

C. Chan, I. O. Moen and J. H. S. Lee, "Influence of Confinement on Flame Acceleration due to Repeated Obstacles", Combustion and Flame, 49 (1983) pp. 27-39.

K. Charles, C. Westbrook and L. C. Haselman, "Chemical Kinetics in LNG Detonations", 7th ICDERS, 1979.

S. Chippett, "Modeling of Vented Deflagrations", Combustion and Flame, 55, (1980) pp. 127-140.

M. Coleman, M. Cain, R. Danna, C. Harley and D. Sharp, "A Review of Energy Release Processes from the Failure of Pneumatic Pressure Vessels", ESMC-TR-88-03, Eastern Space & Missile Center, Patrick AFB, FL, 1988.

P. M. Collins, "Detonation Initiation in Unconfined Fuel-Air Mixtures", Acta Astronautica, 1 (1974) pp. 259-266.

H. F. Coward and G. W. Jones, "Limits of Flammability of Gases and Vapors", Bulletin 503, Bureau of Mines, U.S. Department of the Interior, (1952).

M. Cowperthwaite and W. H. Zwisler, "Tiger Computer Program Documentation", Report No. Z106, Stanford Res. Inst., Menlo Park, CA, 1973.

J. C. Cummings, J. R. Torczynski and W. B. Benedick, "Flame Acceleration in Mixtures of Hydrogen and Air", SAND-86-0173, Sandia Nat. Labs., Albuquerque, NM, 1987.

E. K. Dabora, "The Relation Between Energy and Power for Direct Initiation of Hydrogen-Air Detonations", 2nd Int. Workshop on the Impact of Hydrogen on Water Reactor Safety, 1982a.

E. K. Dabora, "Fundamental Mechanisms of Liquid Spray Detonations", Fuel-Air Explosions, SM Study 16, Univ of Waterloo Press, Ontario, Canada, 1982b.

J. A. Davenport, "A Survey of Vapor Cloud Incidents", Chemical Engineering Progress, 73,9 (1977) pp. 54-63.

B. Deshaies and J. C. Leyer, "Flow Fields Induced by Unconfined Spherical Accelerating Flames", Combustion and Flame, 40 (1981) pp. 141-153.

L. Dirkzwager, "The Effect of Cornstarch Moisture Content on Explosion Ignition of Energy Requirements", Proceedings of the International Symposium on Grain Dust Explosions, Kansas City, MO, (1977).

W. Doering and G. Burkhardt, "Contributions to the Theory of Detonation", Translation from the German as Tech. Report No. F-TS-1227-IA (GDAM A9-T-4G), Headquarters, Air Material Command, Wright-Patterson AFB, (1949).

H. G. Dorsett, M. Jacobson, J. Nagy and R. P. Williams, "RI 5424", U. S. Bureau of Mines, Pittsburgh, PA, 1960.

R. K. Eckhoff, "Use of (dP/dt)max from Closed-Bomb Tests for Predicting Violence of Accidental Dust Explosions in Industrial Plants", First Colloquium on Explosibility of Industrial Dusts, Baranow, Poland, 1984.

R. K. Eckhoff, "Assessing Ignitibility and Explosibility of Dust Clouds in Industry by Laboratory Tests", Proc. of Shengyang Int. Symp. on Dust Explosions, Shengyang, China, 1987.

R. K. Eckhoff, "Scaling of Vented Dust Explosions in Large Silos", Third Int. Coll. on Dust Explosions, Szczyrk, Poland, 1988.

R. K. Eckhoff, K. Fuhre, O. Krest, C. M. Guirao and J. H. S. Lee, "Some Recent Large Scale Gas Explosion Experiments in Norway", Report CMI No. 790750-1, The Christian Michelsen Inst., Fantoft, Norway, 1980.

D. H. Edwards, G. O. Thomas and M. A. Nettleton, "The Diffraction of a Planar Detonation Wave at an Abrupt Area Change", Jour. Fluid Mech., 95 (1979) pp. 79-96.

T. V. Eichler and H. S. Napadensky, "Accidental Vapor Phase Explosions on Transportation Routes Near Nuclear Power Plants", NUREG/CR-0075, (1977).

R. J. Enright, "Experimental Evaluation of the 1.2, 8 and 20 Litre Explosion Chambers", Int. Coll. on Explosibility of Industrial Dusts, Baranow, Poland, 1984.

E. D. Esparza and W. E. Baker, "Measurement of Blast Waves from Bursting Pressurized Frangible Spheres Pressurized with Flash-Evaporating Vapor or Liquid", NASA CR-2811, NASA, Washington, DC, 1977a.

E. D. Esparza and W. E. Baker, "Measurement of Blast Waves from Bursting Pressurized Frangible Spheres", NASA CR-2843, NASA, Wasshington, DC, 1977b.

E. A. Farber, "A Mathematical Model for Defining Explosive Yield and Mixing Probabilities of Liquid Propellant", University of Florida, Gainesville, FL, (1966).

E. A. Farber, J. H. Brann, J. C. Reed, T. M. Reed, G. J. Schoessow, D. Van der Reyden and J. A. Wethington, "Feasibility Study to Explore the Explosive Effects of Liquid Propellants to Define the Mathematical Behavior of Physical Processes Involved", NAS 10-1255, University of Florida, Florida Engineering and Industrial Experiment, Dept. of Mech. Engineering, Gainesville, FL, (1965).

E. A. Farber, J. S. Brann, J. C. Reed, T. M. Reed, G. J. Schoessow, F. L. Schwartz, D. Van der Reyden and J. A. Wethington, "Feasibility Study to Explore the Explosive Effects of Liquid Propellants to Define the Mathematical Behavior of Physical Processes Involved", Interim Report, Univ. of Florida, Florida Engr. and Industrial Experiment Station, Dept. of Mech. Engineering, Gainesville, FL, (1965).

E. A. Farber and J. H. Deese, "A Systematic Approach for the Analytical Analysis and Prediction of the Yield from Liquid Propellant Explosions", University of Florida, Gainesville, FL, (1966).

E. A. Farber, F. W. Klement and C. F. Bonzon, "Prediction of Explosive Yield and Other Characteristics of Liquid Propellant Rocket Explosions", NAS 10-1255, University of Florida, Florida Engineering and Industrial Experiment, Dept. of Mech. Engineering, Gainesville, FL, (1968).

E. A. Farber, R. E. Stewart, C. F. Bonzon and A. Hernandez, "Fireball and Post-Fireball Composition and Atmospheric Chemistry of Fuel/Oxygen-Fluorine Propellants", University of Florida, Florida Engineering and Industrial Experiment Station, Dept. of Mech. Engineering, Gainesville, FL, (1967).

E. A. Farber, et al, "Characteristics of Liquid Rocket Propellant Explosion Phenomena", Technical Paper Series, Florida Engineering and Industrial Experiment Station, College of Engineering, Gainesville, FL, XXI (1967a).

E. A. Farber, et al, "Characteristics of Liquid Rocket Propellant Explosion Phenomena", Technical Paper Series, Florida Engineering and Industrial Experiment Station, College of Engineering, Gainesville, FL, XXI (1967b).

W. Fickett and W. C. Davis, Detonation, Univ. of Cal. Press, 1979.

P. Field, Dust Explosions, Handbook of Powder Technology, Elsevier Scientific Publishing Co., Amsterdam, 1982.

B. Fishburn, N. Slagg and P. Lu, "Blast Effect from a Pancake Shapoed Fuel Drop-Air Cloud Detonation (Theory and Experiment) ", Jour. Haz. Mat., 5 (1981) pp. 65-75.

R. F. Fletcher, "Characteristics of Liquid Propellant Explosions", Annals of the New York Academy of Sciences, 152 (1968) pp. 432-440.

R. S. Fry and J. A. Nicholls, "Blast Initiation and Propagation of Cylindrical Detonations in MAPP-Air Mixtures", AIAA Journal, 12,12 (1974) pp. 1703-1708.

W. G. Garrison, "Large Property Damage Losses in the Hydrocarbon Chemical Industries", M & M Protection Consultants, 1989.

M. Gaug, R. Knystautas, J. H. S. Lee, L. S. Nelson, W. B. Benedick and J. E. Shepherd, "Lean Flammability Limits of Hybrid Mixtures", Dynamics of Reactive Systems. Part II: Modeling and Heterogeneous Combustion, AIAA, New York, NY, 1986.

G. B. Gayle, "Investigation of S-IV All Systems Vehicle Explosion", NASA TM X-53039, NASA George C. Marshall Space Flight Center, Propulsion and Vehicle Engineering Laboratory, Huntsville, AL, (1964).

J. B. Gayle, C. H. Blakewood, J. W. Bransford, W. H. Swindell and R. W. High, "Preliminary Investigation of Blast Hazards of RP-1/LOX and LH 2./LOX Propellant Combinations", NASA TM X-53240, NASA Marshall Flight Center, Research and Dev. Operations, Propulsion and Vehicle Engr. Lab, Huntsville, AL, (1965).

F. C. Gibson, R. W. Watson, J. E. Hay, C. R. Summers and F. H. Scott, "Sensitivity of Propellant Systems", Quarterly Report, Bureau of Mines, U.S. Department of the Interior, (1965).

J. P. Gillis, "New Approaches for Suppression of Dust Fires and Explosions", Shenyang Int. Symp. on Dust Explosions, Shenyang, China, 1987.

I. Glassman, Combustion, Academic Press, 1977.

S. Gordon and B. McBride, "Computer Program for Calculation of Complex Chemical Equilibrium Compositions, Rocket Performance, Incident and Reflected Shocks, and Chapman-Jouguet Detonations", NASA SP-273, NASA, 1976.

K. Gugan, Unconfined Vapor Cloud Explosions, The Institution of Chem. Eng., Ridley Wark, England, 1978.

P. Gunther and G. R. Anderson, "Statistical Analysis of Project Pyro Liquid Propellant Explosion Data", TM-69-1033-3, Bellcom, Inc., Washington, DC, (1969).

F. H. Harlow and A. A. Amsden, "Fluid Dynamics -- An Introductory Text", LA 4100, Los Alamos Scientific Laboratory, University of California, Los Alamos, NM, (1970).

A. J. Harrison and J. A. Eyre, "The Effect of Obstacle Arrays on the Combustion of Large Pre-mixed Gas/Air Clouds", Comb. Sci. Tech., 52 (1986) p. 121.

M. Hertzberg, R. S. Conti and K. L. Cashdollar, "Electrical Ignition Energies and Thermal Autoignition Temperatures for Evaluating Explosion Hazards of Dusts", RI 8988, U. S. Bureau of Mines, Pittsburgh, PA, 1985.

B. H. Hjertager, "Influence of Turbulence on Gas Explosions", The Control and Prevention of Gas Explosions, Oyez Sci. & Tech. Services, London, 1984.

B. H. Hjertager, K. Fuhre, S. J. Parker and J. R. Bakke, "Flame Acceleration of Propane-Air in a Large-Scale Obstructed Tube", Dynamics of Shock Waves, Explosions and Detonations, AIAA, New York, NY, 1984.

J. C. Hokanson and W. O. Seals, "M-1 Propellant Dust Explosibility", Presented at the Conference on Mechanisms of Explosions in Dispersed Energetic Materials, Dover, NJ, (1978).

S. Horiguchi, Y. Urano, K. Tokuhashi, M. Iwasaka and S. Kondo, "Flammability Limits of Silane", Koatsu Gasu, 24 (1987) p. 143.

R. Humbert-Basset and A. Montet, "Dispersion dans l'Atmosphere d'un Nuage Gazeux forme par Epandage de G.N.L. sur le Sol", Third International Conference on Liquefied Natural Gas, Washington, DC, (1972).

K. Kailasanath and E. S. Oran, "The Relation Between Power and Energy in the Shock Initiation of Detonations", NRL Memorandum Report 5179, US Naval Res. Lab., Washington, DC, 1983.

F. Kallis, "Report-Investigation of Blast Effects of Various Fuel Gels Formed in Liquid Oxygen", Experimental Test Memorandum No. 55-55, General Dynamics Corp., Convair Division, San Diego, CA, (1955).

C. W. Kauffman, "Recent Dust Explosion Experiences in the U. S. Grain Industry", Industrial Dust Explosions, ASTM STP 958, ASYM, Philadelphia, PA, 1987.

W. D. Kennedy, "Explosions and Explosives in Air", Effects of Impact and Explosion, Summary Technical Report of Div.2, NDRC, Washington D.C., AD 221 586, I (1946).

C. N. Kingery, "Air Blast Parameters Versus Distance for Hemispherical TNT Surface Bursts", BRL Report No. 1344, Aberdeen Proving Ground, MD, (1966).

C. N. Kingery and B. F. Pannill, "Peak Overpressure Versus Scaled Distance for TNT Surface Bursts (Hemispherical Charges)", Ballistic Research Laboratory, BRL MR 1518, (1964).

A. R. Kiwan, A. L. Arbuckle and L. Giglio-Tos, "Experimental Study of Fuel-Air Detonations at High Altitude", BRL Memorandum Report No. 2554, U. S. Army Ball. Res. Lab., Aberdeen, MD, 1975.

R. Klemens and P. Wolanski, "Flame Structure in Dusted-Air and Hybrid-Air Mixtures, New Lean Flammability Limits", Dynamics of Reactive Systems. Part II: Modeling and Heterogeneous Combustion, AIAA, New York, NY, 1986.

R. Knystautas, C. M. Guirao, J. H. S. Lee and A. Sulmistras, "Measurements of Cell Size in Hydrocarbon-Air Mixtures and Predictions of Critical Tube Diameter. Initiation Energy and Detonability Limits", 9th ICDERS, 1983.

R. Knystautas, J. H. S. Lee, I. Moen and H. G. Wagner, "Direct Initiation of Spherical Detonation by a Hot Turbulent Gas Jet", 17th Int. Comb. Symp., 1979.

S. M. Kogarko, V. V. Aduskin and A. G. Lyamin, "An Investigation of Spherical Detonation of Gas Mixtures", Int. Chem. Eng., 6 (1966) p. 393.

A. Lapin and R. H. Roster, "Oxygen Diffusion in the Atmosphere from Liquid Oxygen Pools", Adv. Cryo. Eng., 13 (1967).

J. H. S. Lee, "Initiation of Gaseous Detonations", Annual Review of Physical Chemistry, 28 (1977) pp. 75-104.

J. H. S. Lee, "Dynamic Parameters of Gaseous Detonations", Ann. Rev. Fluid Mech., 16 (1984a) pp. 311-336.

J. H. S. Lee, "Overview of Gas Explosions and Recent Results in the Study of Turbulent Deflagrations and Detonations", Symp. on Loss Prevention and Safety Promotion in the Process Industries, 1984b.

J. H. S. Lee and C. M. Guirao, SM Study No. 16. Fuel-Air Explosions, University of Waterloo Press, Ontario, Canada, 1982a.

J. H. S. Lee and C. M. Guirao, "Pressure Development in Closed and Vented Vessels", Plant/Operations Progress, 1, 2 (1982b) pp. 75-85.

J. H. S. Lee, C. M. Guirao and D. E. Grierson, "Fuel-Air Explosions", Proceedings of the International Conference on Fuel-Air Explosions held at McGill University, Montreal, Canada, (1981).

J. H. S. Lee, R. Knystautas and C. M. Guirao, "", 15th Symp. Int. Combustion Proc., 1974.

J. H. S. Lee, R. Knystautas and C. M. Guirao, "The Link Between Cell Size, Critical Tube Diameter, Initiation Energy and Detonability Limits", Fuel-Air Explosions, SM Study No. 16, Univ of Waterloo Press, Ontario, Canada, 1982.

J. H. S. Lee and H. Matsui, "A Comparison of the Critical Energies for Direct Initiation of Spherical Detonations", Combustion and Flame, 28 (1977) p. 61.

J. H. S. Lee and K. Ramamurthi, "", Combustion & Flame, 27 (1976) pp. 331-340.

F. P. Lees, Loss Prevention in the Process Industries. Hazard Identification, Assessment and Control, Volumes I and II, Butterworths, London, 1980.

B. Lewis and G. von Elbe, Combustion, Flames and Explosions of Gases, Academic Press, New York, NY, 1961.

J. C. Libouton, M. Dormal and S. J. Van Tiggelen, "Reinitiation Process at the End of a Detonation Cell", Proceedings of Gasdynamics of Explosions, AIAA, New York, NY, 1981.

C. D. Lind, "Unconfined Vapor Cloud Explosions", Loss Prevention, AIChE, 8 (1975a) pp. 18-20.

C. D. Lind, "What Causes Vapor Cloud Explosions?", Loss Prevention, AIChE, 9 (1975b) pp. 101-105.

C. D. Lind and J. C. Whitson, "Explosion Hazards Associated with Spills of Large Quantities of Hazardous Materials, Phase II", Report #CG-D-85-77, U.S. Coast Guard, NTISAD-A047585, (1977).

J. W. Linnett, "Methods of Measuring Burning Velocities", 4th Symp. Int. Comb., 1953.

R. Makepeace, "The Effect of Turbulence on the Combustion of Gases", Fire and Materials, 2 (1978) pp. 85-101.

P. J. Martinkovic, "Space Vehicle Hazards Associated with Leakage of Propellants", Air Force Rocket Propulsion Laboratory, Edwards AFB, CA, (1965).

H. Matsui and K. Komamiya, "An Experimental Study of Soot Film Detonations", Dynamics of Explosions, AIAA, New York, NY, 1986.

H. Matsui and J. H. S. Lee, "", 17th Symp. Int. Comb., 1978.

J. W. Meyer, P. A. Urtiew and A. K. Oppenheim, "On the Inadequacy of Gasdynamic Processes for Triggering to Detonation", Combustion & Flame, 14 (1970) pp. 13-20.

V. V. Mitrofanov and R. I. Soloukin, "The Diffraction of Multifront Detonation Waves", Sov. Phys.-Dokl., 9 (1965) p. 1055.

I. O. Moen, "The Influence of Turbulence on Flame Propagation in Obstacle Environments", Fuel-Air Explosions, SM Study No. 16, Univ. of Waterloo Press, Ontario, Canada, 1982.

I. O. Moen, M. Donato, R. Knystautas and J. H. S. Lee, "The Influence of Confinement on the Propagation of Detonations Near the Detonability Limit", 18th Symp. Int. Comb., 1981.

I. O. Moen, M. Donato, R. Knystautas, J. H. S. Lee and H. G. Wagner, "Turbulent Flame Propagation in the Presence of Obstacles", Gasdynamics of Detonations and Explosions, AIAA, New York, NY, 1981.

I. O. Moen, J. H. S. Lee, B. H. Hjertager, K. Fuhre and R. K. Eckhoff, "Pressure Development due to Turbulent Flame Propagation in Large Scale Methane-Air Explosions", Combustion & Flame, 4 (1982) p. 31.

I. O. Moen, P. Thibault, J. Funk, S. Ward and G. M. Rude, "Detonation Length Scales for Fuel-Air Explosives", 9th ICDERS, 1983.

R. V. Molen and J. A. Nicholls, "Blast Wave Initiation Energy for the Detonation of Methane-Ethane-Air Mixtures", Comb. Sci. & Tech., 21 (1979) pp. 75-78.

B. P. Mullins and S. S. Penner, "Explosions, Detonations, Flammability and Ignition", Agardograph 31, 1959.

J. Nagy and D. W. Mitchell, "Experimental Coal-Dust and Gas Explosions", U.S. Bureau of Mines RI 6344, (1963).

J. Nagy, E. C. Seiler, J. W. Conn and H. C. Verakis, "Explosion Development in Closed Vessels", Bureau of Mines RI 7507, (1971).

H. S. Napadensky and W. W. Bodie, "Safe Siting of Petrochemical Plants", National Symposium on Occupational Safety and Health, Carnegie Institution, Washington, DC, (1973).

S. Ohyagi, T. Yoshihashi and Y. Harigaya, "Direct Initiation of Planar Detonation Waves in Methane/Air/Nitrogen Mixtures", Dynamics of Shock Waves, Explosions and Detonations, AIAA, New York, NY, 1985.

G. Opschoor, "Investigations into the Evaporation of Liquefied Gases Spreading on Land", Cryogenics, 21 (1981) pp. 281-286.

J. J. Oslake, R. J. Getz, R. A. Romine and K. SooHoo, "Explosive Hazards of Rocket Launchings", Technical Report No. U-108:98, Ford Motor Co. Aeroneutronics Division, Ford Road, NewPort Beach, CA, (1960).

K. N. Palmer, Dust Explosions and Fires, Chapman & Hall, London, England, 1973.

J. A. Pappas, "Venting of Large-Scale Volumes", The Control and Prevention of Gas Explosions, Oyez Sci. & Tech. Services, London, 1984.

J. A. Pappas, T. Foyn and D. M. Solberg, "Combustion Instabilities and the Scaling of Vented Gas Deflagrations", 9th International Colloquium on Dynamics of Explosions and Reactive Systems, Poitiers, France, (1983).

D. G. Pearce, D. L. Ward and P. Hayes, "Liquid-Hydrogen Explosions in Containment Vessels", AERE-R-4504, UK Atomic Energy Authority Research Group, Research Reactors Div., Atomic Energy Research Est., Harwell, Berkshire, (1966).

O. Peraldi and B. Veyssiere, "Experimental Study of Detonations in Starch Particle Suspensions with O2/N2, H2/O2 and C2H4/O2 Mixtures", Dynamics of Explosions, AIAA, New York, NY, 1986.

R. E. Pesante, R. D. Erickson, D. G. Frutchey and W. J. Helm, "Blast and Fireball Comparison of Cryogenic and Hypergolic Propellants with Simulated Tankage", Addendum to Report 0822-01(01)FP, Aerojet-General Corp., Research and Engineering Division, Downey, CA, (1964).

R. E. Pesante and M. Nishibayashi, "Evaluation of the Blast Parameters and Fireball Characteristics of Liquid Oxygen/Liquid Hydrogen Propellant", Final Report, 0954-01(01) FP, Aerojet-General Corp., Research Division, Downey, CA, (1967).

J. F. Pittman, "Blast and Fragment Hazards from Bursting Pressure Tanks", NOLTR 72-102, U. S. Naval Ord. Lab., White Oak, MD, 1972.

J. F. Pittman, "Blast and Fragments from Superpressure Vessel Rupture", NSWC/WOL/TR 75-87, NSWC, White Oak, MD, 1976.

M. S. Raju and R. A. Strehlow, "Numerical Investigations of Nonideal Explosions", Jour. Haz. Mat., 9 (1984) pp. 265-290.

A. Rinnan, "Transmission of Detonations Through Tubes and Orifices", Fuel-Air Explosions, SM Study 16, Univ of Waterloo Press, Ontario, Canada, 1982.

A. F. Roberts and D. K. Pritchard, "Blast Effect from Unconfined Vapour Cloud Explosions", Jour. Occ. Acc., 3 (1982) pp. 231-247.

H. C. Rodean, W. J. Hogan, P. A. Urtiew, H. C. Goldwire, T. G. McRae and D. L. Morgan, "Vapor Burn Analysis for the Coyote Series LNG Spill Experiments", UCRL-5350, Lawrence Livermore Nat. Lab., Livermore, CA, 1984.

M. J. Rosenfield, "The Development of Damage Indexes to Structures Due to Liquid Propellant Explosions: Phase II - Damage Indexes", Technical Report No. 4-75, Department of the Army, Ohio River Division Laboratories, Corps of Engineers, Cincinnati, OH, (1968).

R. G. Sachs, "The Dependence of Blast on Ambient Pressure and Temperature", BRL Report No. 466, U. S. Army Ball. Res. Lab., Aberdeen, MD, 1944.

M. J. Sapko, A. L. Furno and J. M. Kuchta, "Flame and Pressure Development of Large-Scale CH4-Air-N2 Explosions", Bureau of Mines RI 8176, (1976).

M. J. Sapko, M. Hertzberg, R. W. Watson and K. L. Cashdollar, "Dust Explosion Size Scaling", Proc. Shengyang Int. Symp. on Dust Explosions, Shenyang, China, 1987.

R. T. Sedgwick and H. R. Kratz, "Fuel Air Explosives: A Parametric Investigation", 10th Symposium on Explosives and Pyrotechnics, Franklin Institute, (1979).

J. A. Senecal, "Explosion Suppression of High Kst Dusts", Flammable Dust Explosion, Int. Bus. Comm., St.Louis, MO, 1988.

M. Sichel, "Recent Development in Gaseous and Heterogeneous Detonations", Eastern Section of the Combustion Institute, 1989.

M. Sichel and J. C. Foster, "The Ground Impulse Generated by a Plane Fuel-Air Explosion with Side Relief", Acta Astranautica, 6 (1979) pp. 243-256.

W. M. Smalley and D. E. Anderson, "The Explosive Potential of Liquid Oxygen and RP-1 Missiles", GM-TR-59-000-00579, Space Technology Laboratories, Inc., Los Angeles, CA, (1959).

L. D. Smoot, W. C. Hecker and G. A. Williams, "Predictions of Propagating Methane Air Flames", Combustion & Flame, 26 (1976) pp. 323-342.

D. M. Solberg and J. A. Pappas, "Modelling of Vented Gas Deflagrations", Paper No. 81-P082, First Specialists Meeting (International) of the Combustion Institute, Bordeaux, France, (1981).

R. A. Strehlow, Fundamentals of Combustion, International Textbook Co., Scranton, PA, 1968a.

R. A. Strehlow, "Gas Phase Detonations: Recent Developments", Combustion & Flame, 12 (1968b) pp. 81-101.

R. A. Strehlow, "Unconfined Vapor Cloud Explosions - An Overview", 14th Symp. Int. Comb., 1973.

R. A. Strehlow, Combustion Fundamentals, McGraw-Hill Book Co., New York, NY, 1984.

R. A. Strehlow and W. E. Baker, "The Characterization of Accidental Explosions", NASA CR-134779, (1975).

R. A. Strehlow and W. E. Baker, "The Characterization and Evaluation of Accidental Explosions", Progress in Energy and Conmbustion Science, 2,1 (1976) pp. 27-60.

R. A. Strehlow and C. D. Engel, "Transverse Waves in Detonation II. Structure and Spacing in H2-O2, C2H2-O2, C2H4-O2 and CH4-O2 Systems", AIAA Jour., 7,3 (1969) p. 492.

R. A. Strehlow, R. T. Luckritz, A. A. Adamczyk and S. A. Shimpi, "The Blast Wave Generated by Spherical Flames", Combustion & Flame, 35 (1979) pp. 297-310.

R. A. Strehlow and R. E. Ricker, "The Blast Wave from a Bursting Sphere", Loss Prevention, AIChE, 10 (1976) pp. 115-121.

D. R. Stull, "Fundamentals of Fire and Explosion", AIChE Monograph Series, 73 (1977).

L. C. Sutherland, "Scaling Law for Estimating Liquid Propellant Explosive Yields", Journal of Spacecraft and Rockets, 15 (1980).

M. M. Swisdak, "Explosion Effects and Properties, Part I-Explosion Effects in Air", Naval Surface Weapons Center, White Oak Laboratory, NSWC/WOL/TR 75-116, (1975).

M. J. Tang, J. A. Nicholls, Z. C. Lin, M. Sichel and C. W. Kauffman, "Direct Initiation of Detonation in a Decane Spray", Dynamics of Explosions, AIAA, New York, NY, 1986.

C. S. Tarifa and P. Perez del Notario, "Some Fundamental Problems on the Combustion of Liquid Oxidizers in Hydrogen", XVII International Astronautical Federation Congress, (1966).

P. H. Taylor, "Recent Research on Vapor Cloud Explosions", Leeds University Short Course on Fire and Explosion, 1987.

S. R. Tieszen, M. P. Sherman, W. B. Benedick and J. E. Shepherd, "Detonation Cell Size Measurements in Hydrogen-Air-Steam Mixtures", Dynamics of Explosions, AIAA, New York, NY, 1982.

G. Tsatsaronis, "Prediction of Propagating Laminar Flames in Methane, Oxygen, Nitrogen Mixtures", Combustion & Flame, 33 (1978) pp. 217-240.

A. C. van den Berg, "The Multi-Energy Method - A Framework for Vapour Cloud Explosion Blast Prediction", Jour. Haz. Mat., 12 (1985) pp. 1-10.

A. C. van den Berg, G. Opschoor and J. P. Zeeuwen, "Blast Effects from Vapor Cloud Explosions", (1987).

A. C. van den Berg, C. J. M. Van Wingerden, J. P. Zeeuwen and H. J. Pasman, "Current Research at TNO on Vapour Cloud Modeling", Int. Conf. on Vapor Cloud Modeling, Boston, MA, 1987.

C. J. M. Van Wingerden, "Experimental Investigation into the Strength of Blast Waves Generated by Vapour Cloud Explosions in Congested Areas", 6th Int. Symp. on Loss Prevention and Safety Promotion in the Process Industries, Oslo, Norway, 1989.

C. J. M. Van Wingerden, A. C. van den Berg and G. Opschoor, "Vapour Cloud Explosion Blast Prediction, Paper No. 9c", Loss Prevention Symposium, AIChE, Houston, TX, 1989.

C. J. M. Van Wingerden, A. C. van den Berg and J. P. Zeeuwen, "Validation of Numerical Codes for the Simulation of Blast Generated by Vapor Cloud Explosions", Dynamics of Explosions, AIAA, New York, NY, 1986.

C. J. M. Van Wingerden and J. P. Zeeuwen, "On the Role of Acoustically Driven Flame Instabilities in Vented Gas Explosions and Their Elimination", Combustion & Flame, 51 (1983a).

C. J. M. Van Wingerden and J. P. Zeeuwen, "Flame Propagation in the Presence of Repeated Obstacles: Influence of Gas Reactivity and Degree of Confinement", Jour. Haz. Mat., 8 (1983b) pp. 138-156.

C. J. M. Van Wingerden and J. P. Zeeuwen, "Investigation of the Explosion-Enhancing Properties of a Pipe-Rack-Like Obstacle Array", Dynamics of Explosions, AIAA, New York, NY, 1986.

A. P. Van Ulden, "On the Spreading of a Heavy Gas Released Near the Ground", Loss Prevention and Safety Promotion in the Process Industries Symposium, Delft, (1974).

R. W. Van Dołah, M. G. Zabetakis, D. S. Burgess and G. S. Scott, "Review of Fire and Explosion Hazards of Flight Vehicle Combustibles", ASD-TR-61-278, U.S. Bureau of Mines, Explosives Research Laboratory, U.S. Department of the Interior, (1961).

E. B. Vanta, J. C. Foster and G. H. Parson, "Detonability of Some Natural Gas-Air Mixtures", AFATL TR-74-80, Eglin AFB, FL, 1974.

B. Veyssiere, "Structure of the Detonations in Gaseous Mixtures Containing Aluminum Particles in Suspension", Dynamics of Explosions, AIAA, New York, NY, 1986.

H. Wagner, "Some Experiments about Flame Acceleration", Fuel-Air Explosions, University of Waterloo Press, Waterloo, Canada, 1982.

C. Westbrook, "Chemical Kinetics of Hydrocarbon Oxidation in Gaseous Detonations", Combustion & Flame, 46 (1982) pp. 191-210.

C. K. Westbrook, "Chemical Kinetics in Gaseous Detonations", Fuel-Air Explosions, SM Study 16, Univ. of Waterloo Press, Ontario, Canada, 1982.

A. H. Wiederman, "Air-Blast and Fragment Environments Produced by the Bursting of Vessels Filled with Very High Pressure Gases", PVP Vol. 106, ASME, New York, NY, 1986.

A. H. Wiedermann, T. V. Eichler and C. A. Kot, "Air Blast Effects on Nuclear Power Plants from Vapor Cloud Explosions", ANL-CT-81-17, Argonne National Laboratory, Argonne, IL, 1981.

B. J. Wiekema, "Vapor Cloud Explosion Model", Jour. Haz. Mat., 3 (1980) pp. 221-232.

B. J. Wiekema, "Vapour Cloud Explosions - An Analysis Based on Accidents", Jour. Haz. Mat., 8 (1984) pp. 295-329.

F. A. Williams, Combustion Theory, Addison Wesley Press, 1985.

A. B. Willoughby, T. C. Goodale and C. Wilton, "Problems Involved in Assessing the Explosion Hazards of Liquid BiPropellant Systems", WSS/CI Paper 64-37, URS Corp., San Mateo, CA, (1964).

A. B. Willoughby, J. Mansfield, T. C. Goodale and C. Wilton, "Summary of Existing Information Concerning the Explosive Potential of the Hypergolic Propellant Conbination N 2.04/50% UDMH-50% N2H4.", AFRPL-TR-65-27, URS Corporation, Burlingame, CA, (1965).

A. B. Willoughby, C. Wilton and J. Mansfield, "Liquid Propellant Explosive Hazards, Volumes I-III", Air Force Rocket Propulsion Laboratory, Edwards, CA, AFRPL-TR-68-92, Final Report, (1968).

J. Wisotski and W. H. Snyer, "Characteristics of Blast Wave Obtained from Cylindrical High Explosive Charges", Denver Research Institute, DRI No. 2286, (1965).

P. Wolanski, "Fundamental Problems of Dust Explosions", Fuel-Air Explosions, Univ of Waterloo Press, 1981.

G. T. Woodward, "Storage and Handling of Liquid Propellants for Rockets and Guided Missiles, 1-C, Progress Report", MIPR No.AF 33(657)-3-R&D-1143, Chemical Process Division, Chemical Research & Development Laboratories, Edgewood Arsenal, MD, (1964).

C. Yao, "Explosion Venting of Low-Strength Equipment and Structures", Loss Prevention, AIChE, New York, 8 (1974).

M. G. Zabetakis, "Flammability Characteristics of Combustible Gases and Vapors", Bulletin 627, Bureau of Mines, U.S. Department of the Interior, (1965).

R. G. Zalosh, "Gas Explosion Tests in Room-Size Vented Enclosures", Loss Prevention, AIChE, New York, 13 (1980) pp. 98-110.

J. P. Zeeuwen, C. J. M. Van Wingerden and R. M. Dauwe, "Experimental Investigation into the Blast Effect Produced by Unconfined Vapour Cloud Explosions", 4th Int. Symp. on Loss Prevention & Safety in the Process Industries, I. Chem. E. Symp. Series No. 80, 1983.

I. B. Zeldovich, S. M. Kogarko and N. N. Simonov, "", Soviet Technical Physics, 1,1 (1956) p. 1689.

APPENDIX A

Selected Closed-Vessel Test Data for Combustible Dusts (NFPA 68)

TABLE A-1. Agricultural Products

Material	Median particle size, μm	Minimum explosive concentration, g/m^3	P_{max}, bar ga	$(dP/dt)_{max}$, bar/sec*	K_{St} bar-m sec	Dust hazard class
Cellulose	33	60	9.7	229	229	2
Cellulose, pulp	42	30	9.9	62	62	1
Cork	42	30	9.6	202	202	2
Corn	28	60	9.4	75	75	1
Egg White	17	125	8.3	38	38	1
Milk, powdered	83	60	5.8	28	28	1
Milk, non-fat, dry	60	-	8.8	125	125	1
Soy Flour	20	200	9.2	110	110	1
Starch, corn	7	-	10.3	202	202	2
Starch, rice	18	60	9.2	101	101	1
Starch, wheat	22	30	9.9	115	115	1
Sugar	30	200	8.5	138	138	1
Sugar, milk	27	60	8.3	82	82	1
Sugar, beet	29	60	8.2	59	59	1
Tapioca	22	125	9.4	62	62	1
Whey	41	125	9.8	140	140	1
Wood Flour	29	-	10.5	205	205	2

* Maximum pressure rise rates measured in cubic meter vessel tests.

TABLE A-2. Carbonaceous Dusts

Material	Median particle size, μm	Minimum explosive concentration, g/m^3	P_{max}, bar ga	$(dP/dt)_{max}$, bar/sec*	K_{St} bar-m sec	Dust hazard class
Charcoal, activated	28	60	7.7	44	44	1
Charcoal, wood	14	60	9.0	10	10	1
Coal, bituminous	24	60	9.2	129	129	1
Coke, petroleum	15	125	7.6	47	47	1
Lampblack	< 10	60	8.4	121	121	1
Lignite	32	60	10.0	151	151	1
Peat, 15% H_2O	-	58	60	10.9	157	1
Peat, 22% H_2O	-	46	125	8.4	69	1
Soot, pine	< 10	-	7.9	26	26	1

* Maximum pressure rise rates measured in cubic meter vessel tests.

TABLE A-3. Chemical Dusts

Material	Median particle size, µm	Minimum explosive concentration, g/m^3	P_{max}, bar ga	$(dP/dt)_{max}$, bar/sec*	K_{St} bar-m sec	Dust hazard class
Adipic Acid	< 10	60	8.0	97	97	1
Anthraquinone	< 10	-	10.6	364	364	3
Ascorbic Acid	39	60	9.0	111	111	1
Calcium Acetate	92	500	5.2	9	9	1
Calcium Acetate	85	250	6.5	21	21	1
Calcium Stearate	12	30	9.1	132	132	1
Carboxymethylcellulose	24	125	9.2	136	136	1
Dextrin	41	60	8.8	106	106	1
Lactose	23	60	7.7	81	81	1
Lead Stearate	12	30	9.2	152	152	1
Methylcellulose	75	60	9.5	134	134	1
Paraformaldehyde	23	60	9.9	178	178	1
Sodium Ascorbate	23	60	8.4	119	119	1
Sodium Stearate	22	30	8.8	123	123	1
Sulfur	20	30	6.8	151	151	1

* Maximum pressure rise rates measured in cubic meter vessel tests.

TABLE A-4. Metal Dusts

Material	Median particle size, µm	Minimum explosive concentration, g/m^3	P_{max}, bar ga	$(dP/dt)_{max}$, bar/sec*	K_{St} bar-m sec	Dust hazard class
Aluminum	29	30	12.4	415	415	3
Bronze	18	750	4.1	31	31	1
Iron Carbonyl	< 10	125	6.1	111	111	1
Magnesium	28	30	17.5	508	508	3
Zinc	10	250	6.7	125	125	1
Zinc	< 10	125	7.3	176	176	1

* Maximum pressure rise rates measured in cubic meter vessel tests.

TABLE A-5. Plastics

Material	Median particle size, μm	Minimum explosive concentration, g/m³	P_{max}, bar ga	$(dP/dt)_{max}$, bar/sec	K_{St} bar-m sec	Dust hazard class
(poly) Acrylamide	10	250	5.9	12	12	1
(poly) Acrylonitrile	25	-	8.5	121	121	1
(poly) Ethylene (Low Pressure Process)	< 10	30	8.0	156	156	1
Epoxy Resin	26	30	7.9	129	129	1
Melamine Resin	18	125	10.2	110	110	1
Melamine, molded (Wood flour and Mineral-filled Phenol-Formaldehyde	15	60	7.5	41	41	1
Melamine, molded (Phenol-Cellulose)	12	60	10.0	127	127	1
(poly) Methyl Acrylate	21	30	9.4	269	269	2
(poly) Methyl Acrylate, Emulsion Polymer	18	30	10.1	202	202	2
Phenolic Resin	< 10	15	9.3	129	129	1
(poly) Propylene	25	30	8.4	101	101	1
Terpene-Phenol Resin	10	15	8.7	143	143	1
Urea-Formaldehyde/ Cellulose, Molded	13	60	10.2	136	136	1
(poly) Vinyl Acetate/ Ethylene Copolymer	32	30	8.6	119	119	1
(poly) Vinyl Alcohol	26	60	8.9	128	128	1
(poly) Vinyl Butyral	65	30	8.9	147	147	1
(poly) Vinyl Chloride	107	200	7.6	46	46	1
(poly) Vinyl Chloride/ Vinyl Acetylene Emulsion Copolymer	35	60	8.2	95	95	1
(poly) Vinyl Chloride/ Ethylene/Vinyl Acetylene Suspension Copolymer	60	60	8.3	98	98	1

APPENDIX B

TABLE B-1. Maximum Oxygen Concentration to Prevent Combustion of Suspensions of Combustible Dust in Air-Carbon Dioxide Mixtures [NFPA 69 (1986)]

Dust	Maximum O_2 Concentration %
Agricultural	
Coffee	17
Cornstarch	11
Dextrin	14
Soy Flour	15
Starch	12
Sucrose	14
Chemical	
Ethylene Diamine Tetra-Acetic Acid	13
Isatoic Anhydride	13
Methionine	15
Ortazol	19
Phenothiazine	17
Phosphorous Pentasulfide	12
Salicylic Acid	17
Sodium Lignosulfate	17
Stearic Acid & Metal Stearates	13
Carbonaceous	
Charcoal	17
Coal, Bituminous	17
Coal, Sub-bituminous	15
Lignite	15
Metal	
Aluminum	2
Antimony	16
Chromium	14
Iron	10
Magnesium	0
Manganese	14
Silicon	12
Thorium	0
Titanium	0
Uranium	0
Vanadium	14
Zinc	10
Zirconium	0
Miscellaneous	
Cellulose	13
Paper	13
Pitch	11
Sewage Sludge	14
Sulfur	12
Wood Flour	16
Plastics Ingredients	
Azelaic Acid	14
Bisphenol A	12
Casein, rennet	17
Hexamethylene tetramine	14
Isophthalic Acid	14
Paraformaldehyde	12
Pentaerythritol	14
Phthalic Anhydride	14
Terephthalic Acid	15
Plastics - Special Resins	
Coumarone-Indene Resin	14
Lignin	17
Phenol, Chlorinated	16
Pinewood Residue	13
Rosin, DK	14
Rubber, Hard	15
Shellac	14
Sodium Resinate	14
Plastics - Thermoplastic Resins	
Acetal	11
Acrylonitrile	13
Butadiene-Styrene	13
Carboxymethyl Cellulose	16
Cellulose Acetate	11
Cellulose Triacetate	12
Cellulose Acetate Butyrate	14
Ethyl Cellulose	11
Methyl Cellulose	13
Methyl Methacrylate	11
Nylon	13
Polycarbonate	15
Polyethylene	12
Polystyrene	14
Polyvinyl Acetate	17
Polyvinyl Butyrate	14
Plastics - Thermosetting Resins	
Allyl Alcohol	13
Dimethyl Isophthalate	13
Dimethyl Terephthalate	12
Epoxy	12
Melamine Formaldehyde	17
Polyethylene Terephthalate	13
Urea Formaldehyde	16

APPENDIX C

Recommendations of NAS Panel on Grain Elevator Explosions [Anon. (1980)]

First Priority Actions

- Continue research on methods for reducing the dust concentration in legs to a level below the lower explosive limit.
- Establish a housekeeping program involving a mechanical dust collection system supplemented by manual or other means.
- Conduct rigorous preventive maintenance, especially on all parts of bucket elevators.
- Use a pre-established and enforced permit procedure whenever welding, cutting, or other open flame work is to be done.
- Incorporate a system to indicate belt slippage and misalignment.
- Incorporate a method to check frequently the temperature and vibration of critical bearings.
- Use devices to extract foreign materials from the incoming grain stream.
- Ground all conveying and electrical equipment.

Second Priority Actions

- Examine the overall functions of mills and elevators to develop a totally new system less subject to the hazards of dust explosions.
- Control dust generation and airborne dust at all grain transfer and discharge points.
- Notify all plant managers that safety is their responsibility. If authority is delegated it must be to an employee who reports directly to the plant manager.
- Apply state-of-the-art techniques to reduce the concentration of airborne dust in and emanating from elevator legs.
- Establish an information center to distribute actively all available information on elevator and mill dust explosions and their causes and prevention.
- Establish a fire and explosion prevention training program at each facility.
- Conduct research to develop economic uses for collected grain dust.
- Locate hammer mills, other grinding equipment, and their dust collection systems separate from the main facility.
- Eliminate all nonessential horizontal surfaces.

- Treat the avoidance of dust explosion hazards as an initial design criteria in the construction of new mills and elevators and the modification of existing structures.
- Continue research on methods for reducing dust concentrations below the lower explosive limit in enclosures other than legs.
- Investigate and report on explosions in a manner that reflects the recommendations made by the panel in its report, "The Investigation of Grain Elevator Explosions," Report NMAB 367-1.

Third Priority Actions

- Follow, to the extent practical, the National Fire Protection Association's standard on explosion venting (No. 68) for all enclosures. Concrete structures should be vented by windows or other openings of the size dictated by this standard.
- Establish a government and industry group to aid in developing and updating explosion prevention regulations for elevators and mills.
- Quantify housekeeping standards for cleanliness in grain-handling facilities that will prevent fires and explosions.
- Coat all nonhorizontal surfaces exposed to airborne dust with a material that will prevent the build-up of layered dust.
- Investigate the effect of electrostatics and absolute humidity on the explosion hazard, including an examination of conveyor belt conductivity and the charging of ungrounded conductive structures.
- Apply state-of-the-art techniques to reduce the concentration of airborne dust below the lower explosive limit where possible in enclosures other than legs.
- If dust is returned to the grain stream do it in the least hazardous manner.
- Use only equipment and installation standards meeting national Electrical Code requirements.

APPENDIX D

TABLE D-1. Fundamental Burning Velocities of Selected Gases (NFPA 678, 1988 ed.)

Gas	Fundamental Burning Velocity, cm/sec	Gas	Fundamental Burning Velocity, cm/sec
Acetone	54	Ethyl acetate	38
Acetylene	166	Ethylene oxide	108
Acrolein	66	Ethylenimine	46
Acrylonitrile	50	n-Heptane	46
Allene (propadiene)	87	Hexadecane	44
Benzene	48	1,5-Hexadiene	52
,n-butyl-	37	n-Hexane	46
,tert.-butyl-	39	1-Hexene	50
,1,2-dimethyl-	37	1-Hexyne	57
,1,2,4-trimethyl-	39	3-Hexyne	53
1,2-Butadiene (methylallene)	68	Hydrogen	312
1,3-Butadiene	64	Isopropyl alcohol	41
,2,3-dimethyl-	52	Isopropylamine	31
,2-methyl-	55	Methane	40
n-Butane	45	,diphenyl-	35
,2-cyclopropyl-	47	Methyl alcohol	56
,2,2-dimethyl-	42	1,2-Pentadiene (ethylallene)	61
,2,3-dimethyl-	43	cis-1,3-Pentadiene	55
,2-methyl-	43	trans-1,3-Pentadiene (piperylene)	54
,2,2,3-trimethyl-	42	,2-methyl-(cis or trans)	46
Butanone	42	1,4-Pentadiene	55
1-Butene	51	2,3-Pentadiene	60
,2-cyclopropyl-	50	n-Pentane	46
,2,3-dimethyl-	46	,2,2-dimethyl-	41
,2-ethyl-	46	,2,3-dimethyl-	43
,2-methyl-	46	,2,4-dimethyl-	42
,3-methyl-	49	,2-methyl-	43
,2,3-dimethyl-2-butene	44	,3-methyl-	43
2-Buten-1-yne (vinylacetylene)	89	,2,2,4-trimethyl-	41
1-Butyne	68	1-Pentene	50
,3,3-dimethyl-	56	,2-methyl-	47
2-Butyne	61	,4-methyl-	48
Carbon disulfide	58	cis-2-Pentene	51
Carbon monoxide	46	1-Pentyne	63
Cyclobutane	67	,4-methyl-	53
,ethyl-	53	2-Pentyne	61
,isopropyl-	46	,4-methyl-	54
,methyl-	52	Propane	46
,methylene-	61	,2-cyclopropyl-	50
Cyclohexane	46	,1-deutero-	40
,methyl-	44	,1-deutero-2-methyl-	40
Cyclopentadiene	46	,2-deutero-2-methyl-	40
Cyclopentane	44	,2,2-dimethyl-	39
,methyl-	42	,2-methyl-	41
Cyclopentene	48	Propene (propylene)	52
Cyclopropane	56	,2-cyclopropyl	53
,cis-1,2-dimethyl-	55	,2-methyl-	44
,trans-1,2-dimethyl-	55	Propionaldehyde	58
,ethyl-	56	Propylene oxide (1,2-epoxypropane)	82
,methyl-	58	1-Propyne	82
,1,1,2-trimethyl-	52	Spiropentane	71
trans-Decalin (decahydronaphthalene)	36	Tetrahydropyran	48
n-Decane	43	Tetralin (tetrahydronaphthalene)	39
1-Decene	44	Toulene (methylbenzene)	41
Diethyl ether	47	Gasoline (100-octane)	40
Dimethyl ether	54	Jet fuel, grade JP-1 (average)	40
Ethane	47	Jet fuel, grade JP-4 (average)	41
Ethene (ethylene)	80		

INDEX